Herausgeber:

Prof. Dr. *A. Davison*   Department of Chemistry, Massachusetts Institute
of Technology, Cambridge, MA 02139, USA

Prof. Dr. *M. J. S. Dewar*   Department of Chemistry, The University of Texas
Austin, TX 7812, USA

Prof. Dr. *K. Hafner*   Institut für Organische Chemie der TH
6100 Darmstadt, Schloßgartenstraße 2

Prof. Dr. *E. Heilbronner*   Physikalisch-Chemisches Institut der Universität
CH-4000 Basel, Klingelbergstraße 80

Prof. Dr. *U. Hofmann*   Institut für Anorganische Chemie der Universität
6900 Heidelberg 1, Tiergartenstraße

Prof. Dr. *K. Niedenzu*   University of Kentucky, College of Arts and Sciences
Department of Chemistry, Lexington, KY 40506, USA

Prof. Dr. *Kl. Schäfer*   Institut für Physikalische Chemie der Universität
6900 Heidelberg 1, Tiergartenstraße

Prof. Dr. *G. Wittig*   Institut für Organische Chemie der Universität
6900 Heidelberg 1, Tiergartenstraße

Schriftleitung:

Dipl.-Chem. *F. Boschke*   Springer-Verlag, 6900 Heidelberg 1, Postfach 1780

Springer-Verlag   6900 Heidelberg 1 · Postfach 1780
Telefon (06221) 49101 · Telex 04-61723
1000 Berlin 33 · Heidelberger Platz 3
Telefon (0311) 822001 · Telex 01-83319

Springer-Verlag   New York, NY 10010 · 175, Fifth Avenue
New York Inc.   Telefon 673-2660

25 Fortschritte der chemischen Forschung
Topics in Current Chemistry

Catalysis

Springer-Verlag
Berlin Heidelberg GmbH 1972

ISBN 978-3-540-05542-6   ISBN 978-3-540-36876-2 (eBook)
DOI 10.1007/978-3-540-36876-2

Contents

The Catalytic Activity of Organic and Metallo-Organic Compounds in Heterogeneous Systems

Dr. Joost Manassen

Plastics Research Laboratory, The Weizmann Institute of Science, Rehovot, Israel

Contents

1

1. Introduction

Several excellent reviews have been written recently about the catalytic activity of organic semiconductors [1,2]. These and also older review articles [3,4] emphasize one particular property of organic materials that have an extended system of conjugated double bonds, namely their semiconductivity.

The purpose of this review is to discuss the catalytic activity of organic and metallo-organic compounds in heterogeneous systems in general. Much material reported in the other review articles will be covered as well, but will be discussed from another point of view. As far as possible the catalytic reaction will be considered in the light of the most recent developments in the study of reaction mechanisms, and as such the catalyst is looked upon as an agent capable of engaging in reversible chemical reactions. This property of the catalyst however may be correlated with physical properties like acidity, redox potential, paramagnetism, semiconductivity, spectral properties etc., the measurement of which can give valuable information about its chemical behaviour.

In order to find our way among the many reactions and catalysts studied, we shall have to classify them into certain prototypes, bearing in mind that such a classification is a limitation in itself.

Class A: *Acid Catalysis*

This is a reaction in which a transfer of hydrogen takes place, either as a proton or as a hydride ion. In the literature the term carbonium ion is often used carelessly to indicate this kind of reaction, without proof of the existence of a reaction intermediate. A more accurate description would be a reaction in which a positive charge is developed at the reaction centre, irrespective of whether the reaction is concerted or involves a free carbonium ion.

Reactions studied include dehydrations of alcohols, double bond shifts in olefins, isomerization of hydrocarbons, racemization of optically active compounds, etc.. In the literature a rather rigid separation is made between a Brønsted acid, which is actually a proton donor, and a Lewis acid, which works as a hydride abstractor. We may illustrate this difference by using the double bond shift in olefins as the model reaction.

(1) Brønsted acid

$$AH + R_1{-}\overset{\text{H}}{\underset{}{C}}{=}\overset{\text{H}}{\underset{}{C}}{-}\overset{\text{H}}{\underset{\text{H}}{C}}{-}R_2 \quad \rightarrow \quad A^- + R_1{-}\overset{\text{H}}{\underset{\text{H}}{C}}{-}\overset{\text{H}}{\underset{+}{C}}{-}\overset{\text{H}}{\underset{\text{H}}{C}}{-}R_2$$

$$R_1{-}\overset{\text{H}}{\underset{\text{H}}{C}}{-}\overset{\text{H}}{\underset{+}{C}}{-}\overset{\text{H}}{\underset{\text{H}}{C}}{-}R_2 + A^- \quad \rightarrow \quad AH + R_1{-}\overset{\text{H}}{\underset{\text{H}}{C}}{-}\overset{\text{H}}{\underset{}{C}}{=}\overset{\text{H}}{\underset{}{C}}{-}R_2$$

(2) Lewis acid

$$L + R_1-\overset{\overset{\displaystyle H}{|}}{C}=\overset{\overset{\displaystyle H}{|}}{C}-\overset{\overset{\displaystyle H}{|}}{\underset{\underset{\displaystyle H}{|}}{C}}-R_2 \quad \rightarrow \quad LH^- + \left[R_1-\overset{\overset{\displaystyle H}{|}}{C}\cdots\overset{\overset{\displaystyle H}{|}}{C}\cdots\overset{\overset{\displaystyle H}{|}}{C}-R_2 \right]^+$$

$$\left[R_1-\overset{\overset{\displaystyle H}{|}}{C}\cdots\overset{\overset{\displaystyle H}{|}}{C}\cdots\overset{\overset{\displaystyle H}{|}}{C}-R_2 \right]^+ + LH^- \quad \rightarrow \quad L + R_1-\overset{\overset{\displaystyle H}{|}}{\underset{\underset{\displaystyle H}{|}}{C}}-\overset{\overset{\displaystyle H}{|}}{C}=\overset{\overset{\displaystyle H}{|}}{C}-R_2$$

For a catalyst to be active in such a reaction, it must be capable of reversibly donating and accepting protons, or reversibly accepting and donating hydride ions. Both cases have been described [5,6].

We wish to stress again that for claritiy's sake the reaction intermediate has been indicated as a carbonium ion, but that a concerned reaction, in which the two steps of addition and abstraction occur simultaneously, is conceivable as well.

The counterpart of this reaction type is *base catalysis*, in which the reaction intermediate would be of the carbanion type. However no such reaction has been described in the literature on organic catalysts.

Class B: *Redox Catalysis*

Under this type of catalysis we may classify those reactions in which reversible electron transfer takes place. The electron to be transferred may be carried by an atom, and in such a case we speak about hydrogen or chlorine transfer and the like. While in the previous case the reaction was characterized by the development of a positive charge at the reactive centre, and we could speak about carbonium ions, in the present case the reaction centre is characterized by the presence of an unpaired spin, and we may speak about free radical reactions, taking into account the same precautions as previously.

The most important representatives of this class of reaction are dehydrogenations and oxidations, and we may illustrate this kind of catalysis by the industrially important dehydrogenation reaction:

$$Cat + R_1-\overset{\overset{\displaystyle H}{|}}{\underset{\underset{\displaystyle H}{|}}{C}}-\overset{\overset{\displaystyle H}{|}}{\underset{\underset{\displaystyle H}{|}}{C}}-R_2 \quad \rightarrow \quad Cat\,H + R_1-\overset{\overset{\displaystyle H}{|}}{\underset{\underset{\displaystyle H}{|}}{C}}\cdot-\overset{\overset{\displaystyle H}{|}}{C}-R_2$$

$$Cat + R_1-\overset{\overset{\displaystyle H}{|}}{C}-\overset{\overset{\displaystyle H}{|}}{\underset{\underset{\displaystyle H}{|}}{C}}-R_2 \quad \rightarrow \quad Cat\,H + R_1-\overset{\overset{\displaystyle H}{|}}{C}=\overset{\overset{\displaystyle H}{|}}{C}-R_2$$

$$2\,Cat\,H \quad \rightarrow \quad 2\,Cat + H_2$$

3

The catalyst oscillates between two oxidation states, Cat and CatH.

Oxidation takes place by the combination of two hydrogen atoms and the formation of molecular hydrogen. In the case of oxidative dehydrogenation, we introduce an oxidizing agent into the reaction mixture and the oxidation step becomes:

$$2 \text{ Cat H} + \text{Ox} \quad \rightarrow \quad 2 \text{ Cat} + \text{H}_2\text{Ox}$$

It is not difficult to envisage an organic molecule as a reversible proton donor/acceptor. Many organic acids containing carboxylic groups, or enolic, phenolic or acetylenic hydrogen are part of common chemical experience.

To envisage an organic molecule as a redox agent may pose more difficulties, but here too simple chemical models are known. The best example of an organic redox catalyst is benzoquinone:

But actually most aromatic molecules may accept an electron and form a ion radical, as is illustrated for the case of naphthalene:

The electron is not localized, but is part of the extended π-electron system [7].

A third type of reaction can be defined, which cannot be classified under the two previous ones, but is important in catalysis:

Class C: *Catalysis of a Symmetry Forbidden Reaction*

When the symmetry properties of reactant and product are different, the reaction generally does not occur and is forbidden, or, to be more realistic, it has an unusually high energy of activation.

If in the elementary step a change of total spin occurs, the reaction is forbidden, e.g. in the *ortho/para* conversion of the hydrogen molecule or the decomposition of N_2O into nitrogen and oxygen (see section on this reaction). Materials containing paramagnetic centres could act as catalysts for this type of reaction, and many examples are actually known.

Another type of symmetry forbidden reaction has been defined by Woodward and Hoffmann [8] and involves a *change in symmetry* of the occupied molecular orbitals during the reaction. Certain materials may act as a catalyst in this case, if they can form a reaction intermediate which either circumvents the symmetry rule [9] or lowers the energy of activation [10].

The division into these three reaction types is necessarily a simplification. In reality matters are much more complicated and many intermediate cases may exist.

In the following pages the principles that have been sketched in this introduction will be applied to experimental studies that have appeared in the literature. The material is classified according to chemical reaction. We have tried to be complete, but offer our apologies in advance to those whose publications we have missed.

2. Activation of Molecular Hydrogen

Quantum theory led in 1927 to the assertion that the hydrogen molecule could exist in two distinct and stable forms, parahydrogen and orthohydrogen. In 1929 parahydrogen was isolated and methods of analysis of mixtures were developed [11,12]. In 1931 the heavy hydrogen isotope deuterium was isolated, and the catalytic chemist could now study reactions which he considered to be of ideal simplicity. The reactions are:

$$p-H_2 \quad \rightleftarrows \quad o-H_2 \tag{1}$$

and

$$H_2 + D_2 \quad \rightleftarrows \quad 2\,HD \tag{2}$$

We would classify reaction (1) in Class C. Reaction (2) on the other hand may be of a polar nature (Class A), if the H–H bond is ruptured heterolytically:

$$H_2 \quad \longrightarrow \quad H^+ + H^-$$

but may be of a free radical nature as well, if the H–H bond is ruptured homolytically (Class B):

$$H_2 \quad \longrightarrow \quad H^{\cdot} + H^{\cdot}$$

The theoretical treatment of reaction (1), when catalyzed by paramagnetic centres, was developed by Wigner [13], and the reaction was shown to occur, using oxygen or nitric oxide in the gas phase, or oxygen, paramagnetic ions or free radicals in solution [14,15]. The interesting aspect of reactions (1) and (2) is that on ortho/para conversion at a paramagnetic centre no H–H bond is broken, and consequently no H/D exchange occurs. On the other hand, when H/D exchange does occur, a hydrogen to hydrogen bond has been broken in the process, and O/P conversion occurs as a consequence. Therefore comparison of the rate of the two processes provides valuable information.

J. Manassen

As activated charcoal, metallo-phthalocyanines and solid free radicals catalyze these reactions, they form the oldest example of heterogeneous catalysis
by organic and metallo-organic compounds.

In Fig. 1, the rate of para to ortho hydrogen conversion over charcoal is
shown as a function of temperature, and is seen to pass through a definite minimum [16-18].

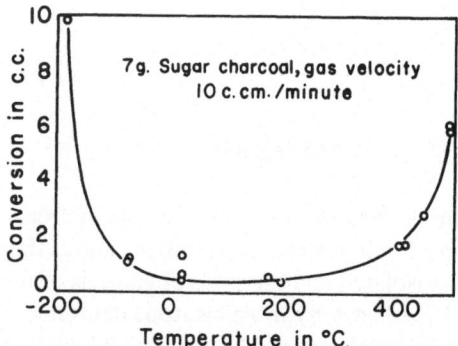

Fig. 1. Rate of ortho/para conversion of hydrogen over dextrose charcoal as a function of
temperature

This minimum was explained by assuming physical adsorbtion in the low
temperature range, which decreases with rising temperature, and chemisorption
in the high temperature range, which increases with temperature.

Copperphthalocyanine and phthalocyanine itself were shown to catalyze
H/D exchange as well as O/P conversion, but deuterium was built into the molecule [19,20]. Later studies showed copperphthalocyanine to be active for O/P conversion, but the metal-free compound did not show reaction for 63 hours at
–80 °C [21]. At working temperatures up to 120 °C no H/D exchange was observed and energies of activation were negative, which is in accordance with the
physical adsorbtion mechanism.

The stable free radical *α,α'-diphenyl-β-picrylhydrazyl* is a paramagnetic
solid and has been studied extensively as a catalyst for the O/P conversion. At
liquid air temperatures no hydrogen adsorption ocurred and conversion was
very slow. Zinc oxide on the other hand did adsorb hydrogen at these temperatures, but did not catalyze the conversion. An intimate mixture of the two gave
a rapid catalytic reaction however [22]. A detailed kinetic study of O/P conversion over the solid free radical in the range 90 – 290 °K showed first order kinetics
and again a negative energy of activation, which was thought to prove the physical adsorbtion mechanism [23]. More modern work confirmed this conclusion,
as it was shown that no H/D exchange occurs, and adsorption of H_2 does not
change the electrical conductivity of the solid, which shows that no chemical
bond with molecular hydrogen is formed [24]. Development of the electron spin

resonance technique made possible the correlation of the activity for O/P conversion with the measured paramagnetism of the solid. Dextrose charcoals, which had been heated at different temperatures, showed a dependence of rate of O/P conversion and D/H exchange on calcination temperature, as shown in Fig. 2 [25].

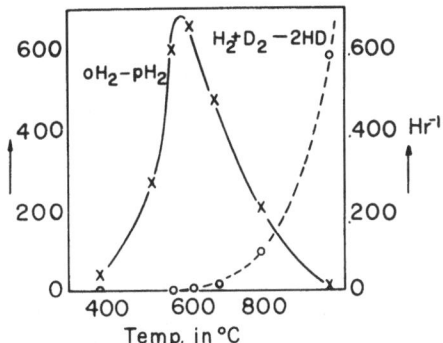

Fig. 2. Rates of O/P conversion and H/D exchange over charcoals as a function of calcination temperature of the catalyst

The sample heated at 605 °C showed the maximum activity for O/P conversion, and this sample also showed the maximum area under the electron spin resonance line and the minimum width, which means the maximum amount of localized paramagnetic centres. The samples heated at higher temperatures showed enhanced activities for H/D exchange and decreasing activities for O/P conversion. The ESR lines were broadened, which indicates exchange interactions between the paramagnetic centres.

In chemical language we might say that the charcoal heated at 605 °C behaves as *a giant free radical*, which, like α,α'-diphenyl-β-picrylhydrazyl, does not bind molecular hydrogen chemically, while the sample heated at 950 °C behaves like a giant olefin, which by reversible hydrogenation/dehydrogenation gives rise to H/D exchange. We shall encounter a comparable phenomenon in the paragraph on N_2O decomposition.

The catalytic activities for O/P conversion and D/H exchange of Cu-phthalocyanine and two of its *polymers* have also been compared [26]. The activation energy for H/D exchange was identical for the three materials, but differences in O/P conversions were found. Measurements were complicated, because the polymers, but not the monomer, tended to adsorb hydrogen in quantities sufficient to add one atom of hydrogen to each copper atom present. That hydrogen was bound to copper could be shown by the disappearance of the copper ESR line after saturation with hydrogen. The fact that this phenomenon occurs shows that hydrogen easily diffused through the entire polymer structure. This and other phenomena to be discussed are an indication that catalytic effects of

7

organic materials cannot always be considered as surface phenomena, and that care has to be taken in expressing catalytic activity per unit surface area. In many cases the *activity per mole of material* may be closer to the truth.

Several pyrolyzed organic polymers could also be shown to be active in the O/P conversion of hydrogen [27].

At the beginning of this section we mentioned that O/P conversion may occur at a paramagnetic centre, while H/D exchange may occur via homolytic or heterolytic splitting of the H—H bond. We may form a paramagnetic as well as polar centre by treating an electron donor with an electron acceptor, forming an electron-donor-acceptor complex. This can be illustrated for the simple case of the reaction between *sodium metal and naphthalene:*

$$\text{Na} + \text{(naphthalene)} \longrightarrow \text{Na}^+ \text{(naphthalene)}^{\cdot -}$$

With a paramagnetic and a polar centre together it is no wonder that these complexes are active catalysts for the activation of molecular hydrogen, and systems of this kind have been studied extensively by Japanese workers. We shall try to summarize the results, some of which have recently been reviewed [28].

If the vapor of *Cs-metal* is brought into contact with layers of *tetracyanopyrene,* a deep violet complex is formed [29,30]. At 77 °K O/P conversion occurs over this complex, but no H/D exchange. Between 190 and 373 °K O/P conversion and H/D exchange occur at comparable rates and with the same activation energy (3.7 Kcal/mole). Neither reaction is catalyzed by Cs-metal or tetracyanopyrene alone. In more extended work of the same nature it was found [31-34] that two mechanisms of exchange can be distinguished:

$$\text{H}_2 + \text{D}_2 \longrightarrow 2\,\text{H D (chemisorption)} \tag{3}$$

$$\text{D}_2 + (\text{Ar})\,\text{H} \longrightarrow \text{HD} + (\text{Ar})\text{D (incorporation} \tag{4}$$
$$\text{of D in complex)}$$

The occurence of mechanism (3) versus (4) depends on the chemical structure of the electron acceptor and also on the ratio between metal and electron-acceptor [35-38]. For instance 1,3-nitrobenzene/sodium does not give mechanism (3), whereas phthalonitrile/sodium does. Naphthacene$^{\cdot -}$/sodium$^+$ reacts mainly according to mechanism (3) (probably a homolytic splitting of H_2), while naphthacene^{2-} (sodium$^+$)$_2$ reacts more according to mechanism (4) (probably heterolytic splitting). Mechanism (3) has a very low energy of activation and is accompanied by changes in semiconductivity. Mechanism (4) shows an appreciable energy of activation and is not accompanied by changes in semiconductivity.

The dependence of the rate of H/D exchange on the electronic properties of the acceptor molecules have been studied extensively [39,40]. Mechanism (4) could be correlated with localization energies, calculated for the aromatic anions in question, and exchange occurred at the sites of highest electron density. Mechanism (3) could be correlated with the electron affinity of the aromatic molecule. The rate of exchange as a function of electron affinity showed a maximum in the intermediate range. Both too high and too low a value for the electron affinity appeared to be unfavourable for the exchange reaction [41,42].

One of the great advantages in studying the catalytic activity of organic solids is that sometimes the heterogeneous reaction may be compared with reactions of the same molecule in solution, by which special solid state effects can be eliminated. An example of this is the study of hydrogen adsorption by *anion radicals and di-anions of anthracene in tetrahydrofuran-solution.* The following mechanisms have been proposed [43]:

$$\text{(5)}$$

anion — radical

$$+ \ 2 \ H^-$$

$$\text{(6)}$$

di — anion

Table 1. *The electronic configuration and activities of various polynegative ions (EDA complexes) of phthalocyanines for the hydrogen exchange reaction*

Complex	Electron configuration	D_2-HZ[a] $(3+\log k)$	H_2-D_2[b] $(3+\log k)$
FePc	d^6	–	–
LiFePc	d^7	–	–
Li_2FePc	d^8	0.08	0.21
Li_3FePC	$d^8 + \pi$	0.56	0.92
Li_4FePc	$d^8 + \pi^2$	1.98	2.08
CoPc	d^7	–	–
LiCoPc	d^8	–	–
Li_2CoPc	$d^8 + \pi$	0.12	0.24
Li_3CoPc	$d^8 + \pi^2$	0.58	0.86
Li_4CoPc	$d^8 + \pi^3$	1.46	1.82
Li_5CoPc	$d^8 + \pi^4$	1.80	2.10
NiPc	d^8	–	–
LiNiPc	$d^8 + \pi$	0.43	0.30
Li_2NiPc	$d^8 + \pi^2$	1.28	1.26
Li_3NiPc	$d^8 + \pi^3$	1.65	1.98
Li_4NiPc	$d^8 + \pi^4$	2.12	2.46

a) $k\,(hr^{-1})$ at $60°$, $P_{D_2} = 12.5$ Hg.
b) $k\,(hr^{-1})$ at $60°$, $P_{H_2} = 15$ cm Hg, $H_2:D_2 = 1:1$.

In our language we could classify mechanism (5) as homolytic splitting (Class B) and mechanism (6) as heterolytic splitting (Class A).

In Table 1 the rate constants for H/D exchange are given according to mechanismus (3) and (4) using as catalysts the electron donor-acceptor complexes between lithium and different metallo-phthalocyanines of known electron configuration [28]. A clear trend can be distinguished: The more electrons are transferred to the π-electronsystem of the phthalocanine ring the more active is the material as a catalyst. These results are reminiscent of the ones shown in Fig. 2, where the charcoals calcined at high temperature, giving broadened ESR-signals, were the best catalysts for H/D exchange.

3. Decomposition of Hydrogen Peroxide

The decomposition of hydrogen peroxide by metal ions of variable valency (Me^z) was studied by Haber and Weiss [44], whose mechanism is still the one accepted today:

$$H_2O_2 + Me^z \longrightarrow OH^- + Me^{z+1} + OH^{\cdot}$$

$$H_2O_2 + Me^{z+1} \longrightarrow H^+ + M^z + HOO^{\cdot}$$

The metal ion is a typical catalyst for a reaction of Class B, but because H^+ and OH^- ions are formed in the process, the reaction may be expected to be pH dependent as well. The catalyst does not necessarily have to be a metal ion. Every substrate that can change oxidation state at the right potential and is capable of reacting with the hydrogen peroxide molecule in its two valency states can be considered a catalyst for this reaction. This means that the best parameter to correlate this reaction with would be the redox potential of the catalyst, which unfortunately is very difficult to measure on a solid material. The best thing to do is to use a catalyst which can be dissolved in a liquid medium of some kind, and to study the redox properties in the dissolved state. Measurements of this kind will be discussed in the section on oxidation and dehydrogenation.

A neat illustration of how the requirements of the chemical reaction define the specificity of the catalyst was given by using *polymeric chelates of bis(hydroxy-8-quinolyl)-4-methane*. The order of reactivity for different complexed metal ions for H_2O_2-decomposition was [45]

$$Co > Mn \gg Cu > Ni > Zn.$$

This order of reactivity may be understood if we consider the ease with which Co and Mn change valency as compared with Cu, Ni and Zn.

By contrast, for N_2O decomposition the order was:

$$Cu \gg Co > Mn > Ni > Zn.$$

Because this is a reaction of Class C, the high activity of the copper complex compared with the other metal ions should probably be explained by its having the right paramagnetic properties for performing the reaction.

Although some weak activity for H_2O_2 decomposition has been described for metallo-phthalocyanines [46], their *polymers* showed much higher activity. Correlations between the catalytic activity and the semiconductivity of the polymer were sought, and high conductivity with a low energy of activation were found to be beneficial[47]. The carriers of the electrical current measured in semiconductivity studies are thought to be free electrons and holes. For these electrons and holes to be created, an electron has to move from one centre in the solid to another, and concepts of ionization energy and electron affinity enter the calculations [4]. Those two concepts are not much different from those of oxidation and reduction and the property of semiconductivity and the oxidation/reduction potential may correlate with the catalytic activity of a solid for a reaction of Class B in a similar fashion.

An interesting effect is the catalytic activity of an *iron phthalocyanine film*, deposited on iron metal, which was shown to have enhanced activity compared with that of metal complex or metal alone [48].

A wide range of *chelate polymers* with different metal ions has been studied as catalysts for H_2O_2 decompositions [49]. Correlations were sought with the K-adsorption edge as measured by X-ray spectroscopy. This property provides information about the electron distribution around the central metal atom and can give much information about its redox properties. The more recently developed technique of electron spectroscopy [50] can give much more accurate information of this kind and may prove to be a powerful tool in this respect.

Several purely *organic polymers* like pyrolized polyacrylonitrile [51], polyaminoquinones [52], vulcanized aniline black or ordinary aniline black [53,54] and pyrolized chlorinated polyvinylchloride [55] were shown to be active for H_2O_2 decomposition. pH dependence was found in some cases, but mostly no correaltions were found with paramagnetism or semiconductivity. It has been shown [56] for the case of *pyrolized polyacrylonitrile*, that this polymer actually behaves as a polyquinone, capable of reversible electron transfer:

As all the organic polymers tried for this reaction may be assumed to contain quinone groups, the most probable explanation for their activity is the capability of groups of this kind to undergo reversible oxidation/reduction.

Therefore, having considered the experimental material published on this reaction, we seen no reason to assume any mechanism other than the well established Haber-Weiss mechanism, which requires a catalyst of the right oxidation/reduction properties.

4. Hydrazine Decomposition

The mechanism of hydrazine decomposition has been studied much less than that of hydrogen peroxide, and the mechanism actually is not known. We may take some clues however from studies of its oxidation in aqueous solution by metal salts, in which kinetic and isotope labelling techniques were used. As the main mechanism for the oxidation of hydrazine by ferric ion the following was proposed [57]:

$$Fe^{3+} + N_2H_4 \quad \longrightarrow \quad Fe^{2+} + N_2H_3 + H^+$$

$$2 N_2H_3 \quad \longrightarrow \quad N_4H_6$$

$$N_4H_6 \quad \longrightarrow \quad 2 NH_3 + N_2$$

This is a typical reaction of Class B, which may be expected to be pH dependent however, because of the formation of a proton during the reaction. The products in this case are ammonia and nitrogen. To convert this mechanism into a catalytic reaction some other step would have to be proposed, in which the ferrous ion is reoxidized into the ferri-ion, but no studies on this have been done up to now.

Hydrazine is also known to decompose into nitrogen and hydrogen over metallic catalysts like platinum or palladium, and the intermediate has been shown by mass spectrometry [58] to be the di-imid N_2H_2:

$$N_2H_4 \quad \xrightarrow{\text{Pt}} \quad H\text{--}N{=}N\text{--}H + H_2$$

This is probably also a reaction of Class B, which however may be expected not to be pH dependent. Therefore, taking into account the scanty evidence available, we may propose that the catalytic activity for this reaction is also dependent on the redox properties of the catalyst, but that the direction of the reaction, either into ammonia and nitrogen or into nitrogen hydrogen respectively, may be connected with acid/base properties.

The most extensive studies on this reaction were done using *copper chelate polymers* of a wide range of ligands as catalysts. The selectivity of the reaction into either ammonia and nitrogen or nitrogen and hydrogen was seen to be dependent on the kind of ligand used [59-61]. The polymers were stated to be more active than the corresponding monomers or even inorganic salts of the same metal ions. We wish to stress again, as in the case of hydrogen adsorption by copperphthalocyanine polymers, that it is dangerous to compare the catalytic activity of an organic polymer with that of an inorganic compound on the basis of specific surface area. In inorganic compounds, containing heavy atoms, the catalytic reactions may be considered to be a surface reaction only. In an organic polymer on the other hand, having an irregular structure and containing light atoms like carbon, hydrogen, nitrogen and oxygen, the bulk of the material may participate in the catalytic reaction, especially in redox reactions. We shall come back to this point when we consider dehydrogenation reactions.

Phthalocyanine complexes, deposited as a film on a metal of the same kind as their central metal ion, appeared to be active for hydrazine decomposition [48]. Wholly organic polymers, prepared by the dechlorination of *poly vinylidene chloride* [62] were also shown to be active, but the ammonia and hydrogen produced reacted with the catalyst, a problem also encountered with the polychelate catalysts.

5. Decomposition of Formic Acid

The decomposition of formic acid has been a popular reaction for studying catalytic behaviour of inorganic as well as organic catalysts. The rate of formation of the products may be followed easily by studying the change of pressure in a closed system. However, the reaction, instead of being simple, is quite complicated from the chemical point of view.

If the catalyst to be studied has acidic sites, formic acid may decompose into carbon-monoxide and water according to a class A mechanism (dehydration). We may assume the mechanism of decomposition to be that of an acid catalyzed decarboxylation [63]:

$$HC\overset{O}{\underset{OH}{\big\langle}} \;+\; H^+ \;\longrightarrow\; HC\overset{O}{\underset{OH_2^+}{\big\langle}} \;\longrightarrow\; H_2O \;+\; HC_+\overset{O}{\big\langle}$$

$$HC_+\overset{O}{\big\langle} \;\longrightarrow\; CO \;+\; H^+$$

Many catalysts however, especially the metallic ones, cause dehydrogenation instead, and the products of reactions are hydrogen and carbon dioxide.

In the many studies that have been done using metallic catalysts, it has been shown that the adsorbed species is a formate ion [64]. So the chemisorbtion step in the dehydrogenation reaction involves a heterolytic splitting of the OH-bond:

$$HC\overset{O}{\underset{OH}{\big\langle}} \;\longrightarrow\; (HC\overset{O}{\underset{O}{\big\langle}})^-_{ads} \;+\; H^+$$

13

J. Manassen

Therefore in cases where this step is rate determining, as could be shown for metals like gold [64], acid/base properties of the catalyst enter into the picture of the dehydrogenation reaction.

It is known from free radical chemistry [65] that the hydrogen attached to a carbonyl carbon is easily abstracted as a hydrogen atom. In aldehydes and formate esters it is this hydrogen that is abstracted by a free radical:

The radical anion $\dot{C}OO^-$, formed by hydrogen abstraction from a formate ion by a OH˙radical has actually been synthesized [66], and was shown to be a strong one-electron reducing agent. Its conjugate acid $\dot{C}OOH$ could also be formed from formic acid, but was somewhat less active as a reducing agent. Therefore we can assume the second step in the dehydrogenation reaction to be the abstraction of a hydrogen atom from the adsorbed formate ion and on metals like nickel this was proved to be the rate determining step [64], this means that in such a case the rate is governed by the redox properties of the catalyst. If then we take into account that many of the catalysts used also catalyze the watergas-reaction

$$H_2O + CO \rightleftharpoons H_2 + CO_2$$

we are confronted with a complicated pattern.

The decomposition of formic acid over different *phthalocyanine complexes* has been studied extensively by Hanke and coworkers [67-70].

In Table 2 the energies of activation for this reaction are given, when different metallo phthalocyanines are used as catalysts [70].

Table 2. *Energies of activation for the decomposition of HCOOH over metal-phthalocyanines (β-Modification) (Average of many measurements)*

Catalyst	ΔE (kcal/Mol)
MnPc	45.5
FePc	24.8
CoPc	24.7
NiPc	24.5
CuPc	22.5
ZnPc	28.2

The manganese complex in particular, but also the zinc complex, shows higher values than the others. They are also the two compounds for which stable formate complexes have been isolated and studied, and we might expect that in these cases the hydrogen abstraction from the formate ion is the rate determining step in the decomposition, as has been discussed in the previous pages. It must be mentioned here that the high value of the activation energy for the manganese complex was not confirmed by other authors [71]. For the other metal ions the rate determining step might be of a different nature, such as the formation of the formate ion or the adsorbtion of the formic acid molecule. The activation energy for the metastable α-modification of Cu and Ni phthalocyanine was found to be several Kcal/mole lower than that for the stable β-modification. This interesting influence of the crystal structure on catalytic activity has not yet been explained.

Several polymers of Ni and Cu-phthalocyanine appeared to be more active than the corresponding monomers, and this was correlated with the thermal activation energy of the electrical conductivity. A complicating factor was that the H_2 to CO_2 ratio was always smaller than one with the polymers, which means that part of the hydrogen was absorbed by the polymer, a phenomenon we have encountered earlier in the section on hydrogen activation.

Hydrogenation of the catalyst during the reaction and subsequent decrease in activity was observed with dehydrochlorinated polyvinylidene chloride [62] and also with Mg-phthalocyanine [72]. It has been pointed out before that on pyrolysis of organic polymers quinonic structures may be formed, and if we consider the Kekulé structure of the phthalocyanine molecule

quinonic
structure

we see that one of the benzene rings does have a quinonic structure. Therefore we may expect that under certain conditions hydrogen is abstracted from the formic acid by these quinonic groups, and that the hydrogen atoms are too firmly bound to be evolved as molecular hydrogen. Phenomena of this kind have been described for pyrolized polyacrylonitrile [56].

6. Oxidations

Oxidations by molecular oxygen have been studied extensively using organic polymers as well as metallo-organic complexes as heterogeneous catalysts. Molecular oxygen is a biradical and most of its oxidation reactions go by a free radical chain mechanism. We may take the oxidation of a hydrocarbon RH, which is of-ten taken as the model reaction for a catalytic reaction, as an example:

Because oxygen does not abstract hydrogen from RH under mild conditions, the reaction has to be initiated by an initiator I

$$I + RH \ \longrightarrow \ IH + R^{\cdot}$$

Free radical $R^{\cdot}$ combines with oxygen to give a peroxyradical

$$R^{\cdot} + O_2 \ \longrightarrow \ ROO^{\cdot}$$

The peroxyradical may abstract a hydrogen atom from RH

$$ROO^{\cdot} + RH \ \longrightarrow \ R^{\cdot} + ROOH$$

by which we have regenerated $R^{\cdot}$ and obtained the hydroperoxide ROOH as the product. This reaction chain may go on until free radicals $R^{\cdot}$ and $ROO^{\cdot}$ do not react with oxygen and RH, but end their lives in a different fashion. Certain molecules like phenols react very well with free radicals, act as chainbreakers and are inhibitors for the oxidation reaction.

The hydroperoxides formed are relatively stable at temperatures below 100 °C, but may be decomposed by almost any agent A, which can complex with it and give a redox reaction, reminiscent of the Haber-Weiss mechanism

$$A + ROOH \ \longrightarrow \ A^+ + RO^{\cdot} + OH^-$$

$$A^+ + ROOH \ \longrightarrow \ A + ROO^{\cdot} + H^+$$

Not only is the hydroperoxide broken down by this reaction, but new free radicals are formed that may start a new chain. Therefore A works as a catalyst for the oxidation reaction.

From this discussion we see that a material may be a catalyst of an oxidation reaction for several reasons:

1. It forms free radicals that may act as chain initiators. Paramagnetic materials can do this for instance.

2. It acts as a redox agent and breaks down hydroperoxides. In the paragraph on H_2O_2 decomposition we have seen that a wide range of materials do this.

3. It removes inhibitors from the reaction mixture. Because inhibitors may make their presence felt at extremely low concentrations this is not an imaginery possibility.

4. A fourth possibility is that of oxygen activation, which means that the chemical properties of the oxygen molecule are changed by complexation with the catalyst. This activation may take several forms:

a. *Oxygen Carrier*

$$C + O_2 \ \longrightarrow \ C \ldots O_2$$

the bond between oxygen and catalyst is weak and the function of the catalyst is to carry the oxygen. The best example is that of haemoglobin in the blood. Oxygen reacts as molecular oxygen, but its solubility is enhanced by specific solvation with the carrier.

b. *Charge Transfer*

$$C + O_2 \longrightarrow C^{\delta+} \ldots O_2^{\delta-}$$

By partial charge transfer from the catalyst to the oxygen, the latter obtains properties different from those of molecular oxygen and is chemically activated. This mechanism has often been proposed but has never actually been proven.

c. *Electron Transfer*

$$C + O_2 \longrightarrow C^+ + O_2^-$$

By electron transfer to oxygen the catalyst is transformed into its oxydized species and oxygen into an anion radical. Both may initiate or carry chains by abstracting hydrogen from the substrate.

5. The fifth possibility is that of a chain carrier, which can best be illustrated for the case of HBr

$$ROO^\cdot + HBr \longrightarrow ROOH + Br^\cdot$$

$$BR^\cdot + RH \longrightarrow HBr + R^\cdot$$

If the reaction between the peroxyradical and RH is sluggish, its rate may be enhanced because hydrogen is abstracted from HBr, after which the bromine atom formed in its turn abstracts a hydrogen atom from RH.

In spite of the fact that this classification makes things look quite complicated, it is a dangerously simplified one, and the real situation is much more involved. Therefore it is not surprising that many organic materials have been reported as catalysts for oxidation, but that the explanations given for their activity are often contradictory. *Pyrolized polyacrylonitrile* and *polyphenylacetylene* are reported to inhibit the oxidation of cumene [56]. This may be connected with the reported quinonic structure of these polymers, which makes them active towards free radicals.

On the other hand *dehydrochlorinated polyvinylchloride* [73] and *polimethyl-β-chlorvinyl-ketone* [74] catalyze the autoxidation of hydrocarbons, and the activities are related to the semiconductive properties of the catalysts. Recently it has been shown that entirely inert polymers like *polyethylene, polypropylene* and *poly(tetrafluoro) ethylene* are rather efficient catalysts for the oxidation of tetralin [75].

However most work has been done using metallo-organic complexes as catalysts, and the mechanistic picture here is somewhat brighter. In this review we

17

are concerned only with heterogeneous catalysts, most of which were phthalocyanine complexes, which we shall review in somewhat more detail.

Hock and Kropf compared the catalytic activities of different metal complexes of phthalocyanine for the *oxidation of cumene,* and their results are recorded
in Table 3.

Table 3. *Oxidation of cumene, catalyzed by metal phthalocyanine at 80 °C for ten hours*

Catalyst	Mn	Fe^{2+}	Co	Ni	Cu	Zn	Mg
Hydroperoxide %	17.9	12.4	5.8	31.7	12.9	14.2	10.3
Hydrocarbon converted %	63.1	70.3	77.1	42.2	12.9	14.2	10.3
% Hydroperoxide in product	28.4	17.6	7.5	75	>98	>98	>98

Mn, Fe, and Co-phthalocyanines appear to be good catalysts for the oxidation,
but also for hydroperoxide breakdown and relatively small quantities of hydroperoxide are found in the reaction product. Cu, Zn and Mg-phthalocyanines are
also reasonable catalysts for the autoxidation, but in this case most of the hydroperoxide is found undecomposed as the reaction product. In the case of Ni only
part of the hydroperoxide is found intact. In subsequent kinetic studies [77,78]
Kropf decided that we have here a case of oxygen activation, when the reaction
is performed below 100 °C

$$MePc + O_2 \longrightarrow MePc^{\delta+}O_2^{\delta-}$$

The charge transfer complex initiates reaction chains with the formation of free
radicals. At temperatures above 105 °C we get a change in mechanism and hydroperoxide decomposition becomes predominant. As additional proof for this mechanism, it was found that with vanadyl-phthalocyanine, where there is no possibility of complexation with oxygen, no such transition occurs [79]. When substitu-
ents are introduced on the phthalocyanine ring, their influence on the catalytic
activity follow Tafts σ_R constants in such a way, that electron-donating substi-
tuents enhance the rate, while electron-withdrawing groups retard it [80]. This is
also in accordance with the oxygen activation mechanism.

Russian workers came to more or less identical conclusions [81,82], but also showed that a change from the metastable α-modification into the stable β-modifica-
tion changes catalytic activity, a phenomenon also observed for HCOOH decomposition. It has been stressed before that one of the advantages of the study of
heterogeneous organic catalysts is that the catalyst may also be studied in solution. This was done for the oxidation of cysteine using ordinary phthalocyanine
complexes as heterogeneous catalysts and the tetrasulfonated ones as homogeneous catalysts. The rates followed the same trends as a function of composition,
and the homogeneous and heterogeneous reactions probably follow the same
mechanism [83].

In a more quantitative study the redox properties of the phthalocyanine complex and the structurally similar *tetraphenylporphyrins* were studied by electrochemical methods in solution [84]. It was found that the oxidation/reduction may go by two distinctly different mechanisms:

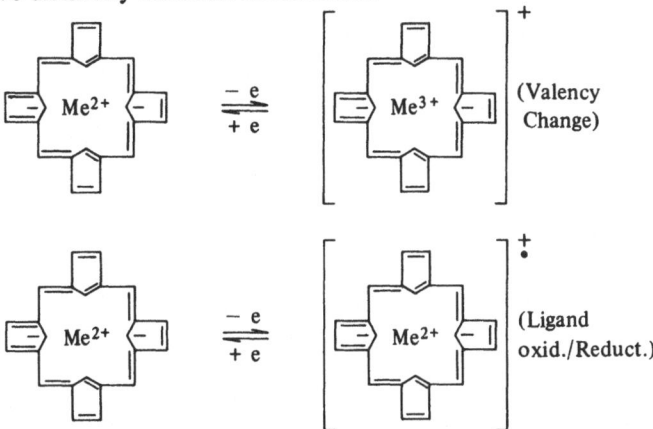

Fe and Co-phthalocyanine appeared to react according to the valency change mechanism, while Cu and Zn-phthalocyanine reacted according to the ligand oxidation/reduction mechanism. The Ni complex appeared to be capable of giving both mechanisms.

The results of Table 3 become clear when we consider these results. The complexes that react according to a valency change mechanism act as catalysts because of hydroperoxide decomposition, while in the case of zinc and copper another mechanism clearly operates, and we propose here that in this case the complex acts as a chain carrier, much like the example of HBr at the beginning of this section (for simplicity's sake the phthalocyanine π-electron system is indicated as a square):

$$\boxed{Me^{2+}}^{+\cdot} + RH \longrightarrow \boxed{Me^{2+}}^{+}{-}H + R^{\cdot}$$

$$R^{\cdot} + O_2 \longrightarrow ROO^{\cdot}$$

$$ROO^{\cdot} + \boxed{Me^{2+}}^{+}{-}H \longrightarrow ROOH + \boxed{Me^{2+}}^{+}$$

The ion radical

$$\boxed{Me^{2+}}^{+\cdot}$$

could be shown by electrochemical methods to be a rather stable species, not less than the bromine atom, while the conjugate acid

$$\boxed{Me^{2+}} \overset{+}{-}H$$

is known to exist in acidic solutions of these complexes [85]. In this connection it is worthwhile to mention that the oxidation of cysteine under the catalytic influence of metallophthalocyanines has been shown to be pH dependent [83]. In the case of Ni Pc both mechanisms possibly operate.

To check the oxygen activation mechanism the comlex $(CuPcO)_2$ has recently been prepared and its properties as a catalyst compared with those of ordinary Cu-Phthalocyanine [86]. It was found that $(CuPcO)_2$ was superior for the oxidation of α-methylstyrene, which occurs by an addition mechanism, but for the ordinary abstraction mechanism in the oxidation of cumene the copper phthalocyanine was superior.

Interesting results have been obtained in studies of the catalytic activity for oxidation by phthalocyanine polymers, containing different metal ions in the same molecule [87-90]. If Fe was mixed with a series of other transition metal ions, differences in activity were found to be dependent on the metal ion, and correlations between the catalytic activity and the thermal activation energy of semiconductivity were found. With copper as the second metal ion, maximum activities were found at a ratio Fe/Cu = 1. Many other chelate polymers have been tested for their oxidation activity, and a dependence of the catalytic activity on the donor properties of the ligand was found [91,92].

Phthalocyanine complexes are some of the few heterogeneous organic catalysts that have found their way into *practical application*. They are used to remove mercaptans from oil by selective oxidation into disulfides [98] and they also used in fuel cell cathodes, where they catalyze the reduction of oxygen [94].

7. Dehydrogenation

In the introduction we discussed how in dehydrogenation we may expect the catalyst to be reduced by abstraction of hydrogen atoms from the substrate, whereupon molecular hydrogen is formed and the catalyst returns to its original oxidation state. If the hydrogen-to-catalyst bond is too strong, the catalyst is hydrogenated during the reaction, with subsequent decrease in activity. We encountered this phenomenon while discussing the decomposition reaction of hydrazine and formic acid, and it also occurs on dehydrogenation of different substrates by pyrolized polyacrylonitrile [56]. By mixing the substrate with an oxidizing agent, this phenomenon may be prevented and we get oxidative dehydrogenation.

Dehydrogenation with the formation of molecular hydrogen has also been reported occasionally for substrates other than hydrazine or formic acid, over organic catalysts, and a patent has been granted for the use of *polyacrylonitrile*, pyrolized on an alumina support for this purpose [95].

In principle one might expect that a catalyst containing Lewis acid sites would also cause dehydrogenation according to a Class A mechanism:

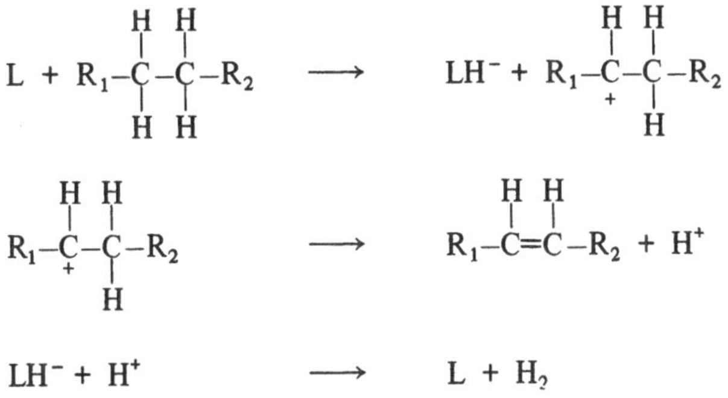

$$L + R_1-\underset{\underset{H}{|}}{\overset{\overset{H}{|}}{C}}-\underset{\underset{H}{|}}{\overset{\overset{H}{|}}{C}}-R_2 \longrightarrow LH^- + R_1-\underset{\overset{|}{+}}{\overset{\overset{H}{|}}{C}}-\underset{\underset{H}{|}}{\overset{\overset{H}{|}}{C}}-R_2$$

$$R_1-\underset{\overset{|}{+}}{\overset{\overset{H}{|}}{C}}-\underset{\underset{H}{|}}{\overset{\overset{H}{|}}{C}}-R_2 \longrightarrow R_1-\overset{\overset{H}{|}}{C}=\overset{\overset{H}{|}}{C}-R_2 + H^+$$

$$LH^- + H^+ \longrightarrow L + H_2$$

A substrate, 5-ethyl-5-methyl-1,3-cyclohexadiene, has been proposed to distinguish this mechanism from the Class B mechanism [56]:

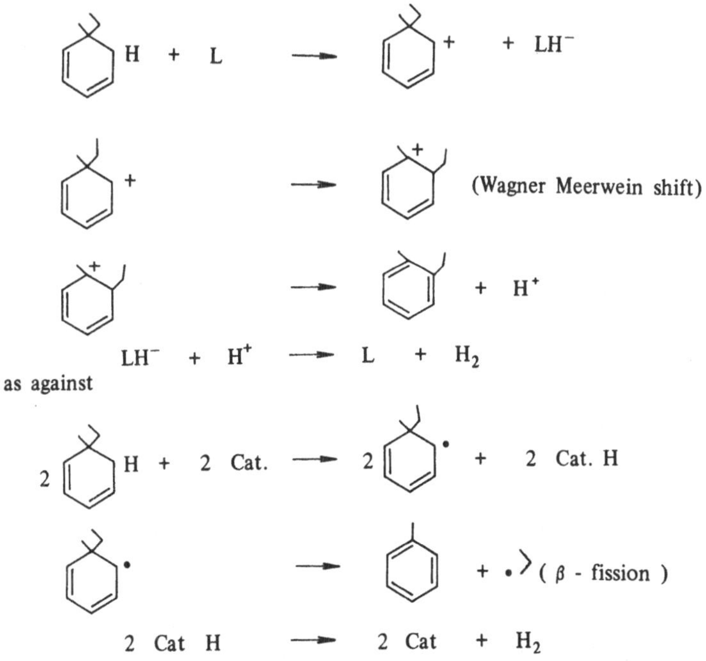

(Wagner Meerwein shift)

$$LH^- + H^+ \longrightarrow L + H_2$$

as against

(β - fission)

$$2 \ Cat \ H \longrightarrow 2 \ Cat + H_2$$

Because acid-catalyzed reactions are prone to give Wagner-Meerwein shifts, we may expect the rearranged reaction product methyl-ethyl-benzene with an acidic catalyst. Free radicals, on the other hand, do not generally give 1,2-shifts, but tend to decompose by β-fission, which gives toluene as the reaction product. It was shown that acidic alumina gives mostly the rearranged product,

21

while classical dehydrogenation catalysts like chromia-alumina or iron-oxide give toluene.

However most mechanistic studies using organic or metallo-organic catalysts have been done on oxidative dehydrogenation according to a redox mechanism, and we shall give a summary of the results:

To prove that the *quinonic structure* is responsible for dehydrogenation activity in organic polymers, two polymers were compared, similar in structure, but one containing quinonic groups and the other without [96].

(contains quinonic groups)

(does not contain quinonic groups)

The first polymer was shown to be active for the dehydrogenation of several substrates, while the second one was entirely inactive under the same conditions. A free radical mechanism could be proven, using the substrate 5-ethyl-5-methyl-1,3-cyclohexadiene.

By introducing benzoquinone structures into a polymer, activities for dehydrogenation were obtained which were comparable to those of commercial dehydrogenation catalysts. An example is the polyquinone [97]:

Another polyquinone, which can be synthesized by addition polymerization of diazotized benzidine with benzoquinone [98] also appeared to be active.

This polymer is red in its oxidized form and yellow in its reduced form, and lends itself to spectroscopic study. Its activity was compared with that of molybdate catalysts, which are well known to be active for oxidative dehydrogenation [99], and many parallels between the organic and inorganic materials were found. The color changes of this catalyst during reaction made it clear that we are not concerned with a surface reaction, but that at least a part of the bulk of the material participates in the oxidative dehydrogenation reaction, a phenomenon we have mentioned several times in these pages.

Most of these organic polymers are insoluble, infusible substances, and to find a more quantitative relation between the redox properties of the catalyst and its catalytic activity, the need arose for a catalyst which could be studied heterogeneously, but whose redox properties may also be studied in solution. Most suitable for this purpose appeared to be the metal complexes of the phthalocyanines and tetraphenylporphyrins. These molecules may be considered as hydroquinonic structures, which becomes clear on comparison of the following two sets of equilibria:

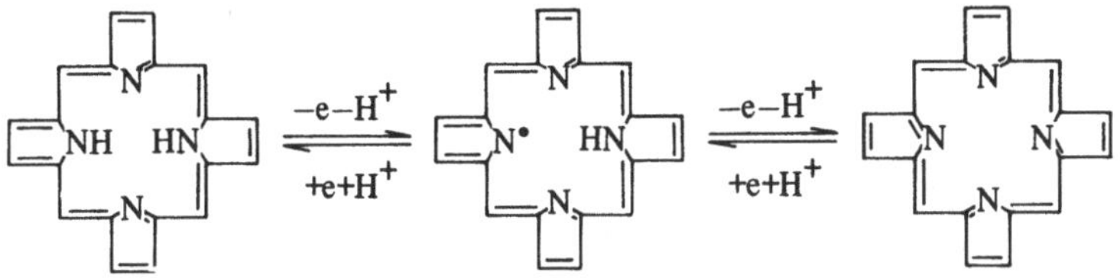

18 π - electrons   semiquinonic   16 π - electrons
aromatic (4n + 2)   quinonic

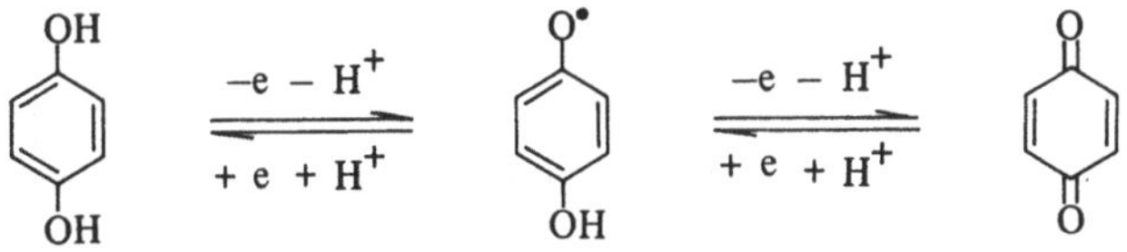

aromatic   semiquinone   quinone

The 18 π-electron structure has the stability of an aromatic ring according to the Hückel $4n + 2$ rule, while the others may be considered to be active like the semiquinone or quinone. So, when an oxidizing agent can be found to oxidize the aromatic molecule, a structure is formed which may abstract a hydrogen atom, by which it is reduced again. A mixture of nitrobenzene and cyclohexadiene-1,4 suited this purpose, and complexes of phthalocyanine and tetraphenylporphyrin were found to catalyze the oxidative dehydrogenation of cyclohexadiene by nitrobenzene [100]:

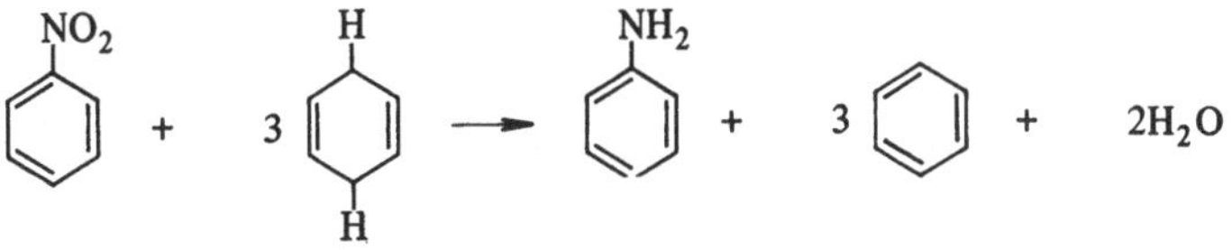

By complexing phthalocyanine or tetraphenylporphyrin molecules with different bivalent metal ions, their oxidation potential may be changed, and this also appeared to change their catalytic activity. In Fig. 3 this is shown graphically.

We see that the correlation is semiquantitative. The higher the potential of the complex, i.e. the more difficult it is to remove an electron from it, the less ac-

23

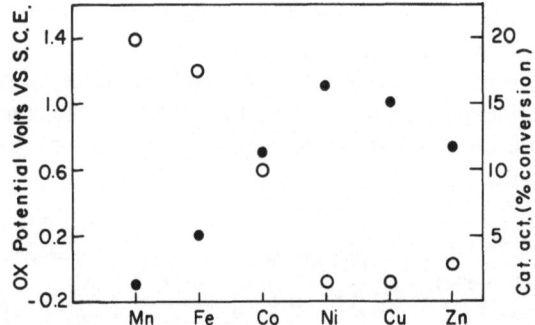

Fig. 3. Oxidation potential and catalytic activity of different metalphthalocyanines as a function of the metal ion. ● Oxidation potential. ○ Catalytic activity

tive it is as a catalyst, in accordance with the proposed mechanism. In the section on oxidations we mentioned that Fe and Co phthalocyanine give a redox reaction by valency change, while Cu and Zn react by ligand oxidation/reduction. From the graph we see that Co and Zn phthalocyanine have approximately the same oxidation potential, but that the Zn complex is the less catalytically active one. Therefore it seems that ligand oxidation/reduction is less effective for the oxidative dehydrogenation of cyclohexadiene than valency change.

Oxidation potentials measure the energy required to remove an electron from a molecule. To check whether in a partial electron transfer in charge transfer complex formation, a similar correlation would hold, the equilibrium constant

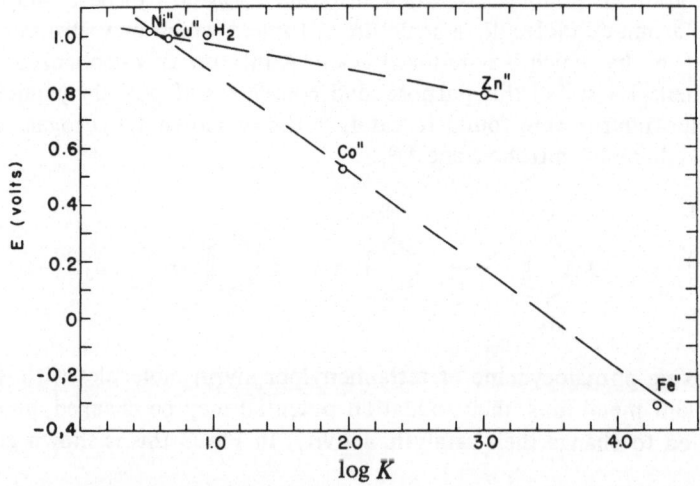

Fig. 4. Relation between oxidation potential and the logarithm of the equilibrium constant of charge transfer complex formation for different metaltetraphenylporphyrins

of charge transfer complex formation between tetraphenylporphyrin complexes and trinitrobenzene was measured by NMR methods in solution [101].

$$\text{Metaltetraphenylporphyrin} + \text{Trinitrobenzene} \underset{}{\overset{K}{\rightleftarrows}} \text{Complex}$$

In Fig. 4 a realtionship between the logarithm of this equilibrium constant and the oxidation potential is shown for different tetraphenylporphyrin complexes, and we see that those complexes which give a redox reaction by valency change, are related differently to this equilibrium constant from the ones reacting by ligand oxidation/reduction. This seems to indicate that in partial electron transfer too these two mechanisms may be distinguished. We encountered a similar phenomenon in the section on hydrogen activation (Table 1), where it was shown that only those electron donor/acceptor complexes which gave ligand reduction were active in H/D exchange.

8. Nitrous Oxide Decomposition

The nitrous oxide molecule can be considered to be a resonance hybrid between the structures:

$$N\equiv N^+\!\!-O^- \qquad \text{and} \qquad N\overset{-}{=}N\overset{+}{=}O$$

It is a reasonably stable molecule, which decomposes by a unimolecular process at temperatures above 600 °C into nitrogen and oxygen. If we consider the energetics of the elementary process in this reaction:

$$N_2O \longrightarrow N_2 + O(^3P) \qquad \Delta H° = 39.7 \text{ Kcal/mole}$$

$$N_2O \longrightarrow N^2 + O(^1D) \qquad \Delta H° = 85.1 \text{ Kcal/mole}$$

we see that the reaction involving the lowest change in energy is associated with a change in total spin [102]. So this is an example of a Class C reaction.

We might expect catalytic action by materials containing paramagnetic centres, and interesting correlations have indeed been found between the electromagnetic properties of organic polymers and their catalytic activity for this reaction[103-106]. Two classes of polymers were studied, the *polypyromellitimides,*

and the *polyindoloimidazoles*

By changing the group Ar it was possible to build into the polymer stretches of conjugated double bonds of different length, or even to build polymers with an uninterrupted conjugated chain. Several correlations were found:

1. Those polymers having an uninterrupted conjugated chain of double bonds, showed a distinctly lower energy of activation for the reaction than the ones in which conjugation was interrupted periodically (20–25 Kcal/mole versus 40–50 Kcal/mole).

2. No direct correlation existed between the number of free spins and the catalytic activity of the free spins, called "A". It was shown however, that in the same family of polymers, A depends linearly on the lenght of the uninterrupted stretch of conjugated double bonds.

3. In the polymers containing an uninterrupted chain of conjugated double bonds the value of A was higher, the more heteroatoms were built into the molecule, and a semiquantitative relationship could be found between the parameter $C/(C+O+N)$ and log A.

4. By subjecting a certain polymer to heat treatment at different temperatures, the number of spins changed, and with it the catalytic activity. It was shown that A correlated well with the width of the ESR line and with the relaxation time T_2. This correlation between A and relaxation time did not hold only for the pyrolized polymers, but for all the polymers studied. This effect was explained by correlating the relaxation time with exchange interactions. The more extensive the exchange interactions (identified as "graphitization"), the weaker the catalytic activity of the free spins. This is in accordance with the absence of catalytic activity of graphite itself. In this way, also the activating effect of heteroatoms, which are known to delay graphitization, was explained. In the less active polymers small graphitized domains were identified by X-ray diffraction.

Correlations of this kind are reminiscent of those pictured in Fig. 2 for the O/P-conversion of hydrogen, and it is clear that by the proper use of organic model polymers, interesting correaltions between their structure, paramagnetic properties and catalytic activity for Class C reactions can be found.

9. Acid Catalyzed Reactions

Typical acid-catalyzed reactions like the dehydration of alcohols and double bond shifts in olefins have been mentioned occasionally as reactions catalyzed by organic heterogeneous catalysts. An extensive kinetic study of the dehydration of tertiary butyl alcohol over pyrolized polyacrylonitrile has been described [107]. In most of these cases the acidity of the catalyst was not studied however, and with substrates like tertiary butanol or dimethylvinylcarbinol [108], that cannot undergo dehydrogenation, the mechanism is not always clear. On the other hand, many acid-catalyzed reactions in which cation exchangers are used as the catalysts, are mentioned in the patent literature and appear in the handbooks dealing with ion exchangers. We shall be concerned in this section only with those studies that deal with acid-catalyzed reactions by heterogeneous organic catalysts, in which the acidity of the catalyst was known.

Instead of the conventional gel-like structures, that swell in contact with a solvent, cation exchangers may be synthesized as a macroreticular polymer. By performing the polymerization in a solvent in the presence of a high percentage of a crosslinking agent, a rigid macroporous structure is obtained, much like a conventional inorganic catalyst support. A detailed study of the catalytic activity of a *sulfonic acid resin* of this type has been published for the racemization of an optically active ketone. The conclusion was drawn [109] that under the reaction conditions only a surface reaction occurred, and that the functional groups within the polymer do not participate in the reaction.

Conventional ion exchange resins are not very thermostable, and for reaction temperatures above 150 °C other synthetic methods have to be tried.

Benzene may be polymerized under the action of aluminium chloride and copper chloride into a thermostable structure which retains the chemical reactivity of benzene. Such a polymer may be sulfonated or phosphonated in suspension, and active acidic catalysts are obtained that are stable up to 350 °C and carry the functional groups only at the surface [110]:

The functional groups on this catalyst may be titrated, the catalytic activity regulated by partial neutralization, and suppressed entirely by total neutralization.

Of the two isomeric alcohols *menthol and neomenthol,* the former dehydrates preferably into menthene-2, while the latter gives menthene-3 preferably.

menthol   menthene-2   neomenthol   menthene-3

This specifity is dependent on the acidity of the catalyst, and was studied using the heterogeneous catalysts sulfonated and phosphonated polybenzene, and the homogeneous catalysts paratoluenesulfonic acid and phosphoric acid. The differences between the sulfonic and phosphonic acid groups were greater than the differences between heterogeneous and homogeneous phase, in spite of a considerable difference in reaction temperature. It was thus shown that for a typical Class A reaction, the type of acidity is more important than the physical state of the catalyst. These acidic catalysts were extremely active, and had to be partially neutralized in order to obtain reasonable rates ans selective reactions.

The isomerization of *butene-1* and *butene-2* has been studied over conventional ion exchange resins [110,111] as well as over the sulfonated and phosponated polyphenyls.[110]. A careful kinetic study showed that the reaction over these materials as well as over silica-alumina catalysts goes by way of a common intermediate, which can be understood best as the secondary butyl carbonium ion.

Several kinetic studies of alcohol and formic acid-dehydration have been described, using cation exchange resins, and the results led to postulates about the possible structure of the adsorbed molecule and the mechanism of its decomposition [112-114].

10. Miscellaneous Reactions

In the last section we described catalysis by cation exchangers for acid-catalyzed reactions. If polymers are made, containing redox groups instead of acidic groups, materials are obtained which are known as electron exchangers. Some catalytic applications of these resins are described in the textbook about redox polymers [115]. The main disadvantage of these electron exchangers, when used as catalysts, is their low thermostability, because of which only reactions that occur at low temperatures may be tested. The quinone group-containing organic catalysts, which have been described in the section on dehydrogenation, are actually thermostable equivalents of this class of polymers.

Chelating resins, that change valency at a desired potential have also been described as catalysts for redox reactions [116]. A quaternary ammonium ion exchange resin treated with a solution of sodium tungstate appeared to be a good heterogeneous catalyst for the epoxidation of maleic acid [117].

A related development is that of polymeric ligands, examples of which have started to appear in the patent literature [118-120]. Since the development of homogeneous catalysis, by which the influence of the ligand on the catalytic activity of a dissolved metal ion was dramatically shown, attempts have been made to attach these ligands to a polymer, and to make the homogenous catalyst insoluble. Thereby one retains the advantage of the homogeneous catalyst, without its inherent disadvantage of having catalyst and reaction mixture in one phase. In most cases a metal salt is treated with an existing or modified ion exchange resin, and heterogeneous catalysts are obtained. In Table 4 a summary is given of what resin/metal combinations were shown to be active as catalysts for different reaction types:

Table 4. *Different reaction types catalyzed by materials, made from a metal salt and a polymeric ligand*

Reaction	Metal salt	Resin	Ref.
Carbonylation	$Pd(NH_3)_4Cl_2$	Amberlyst 15	118
Carbonylation	K_2PdCl_4	Amberlyst A 21	118
Carbonylation	Bis(benzonitrile)-dichloropalladium (II)	Chloromethylated polystyren-eresin reacted with chlorodi-phenylphosphine	118
Carbonylation	Rh or Ru-acetate in HBF_4 + phosphine	Not specified	121
Hydrogenation	Rh or Ru-acetate in HBF_4 + phosphine	Not specified	121
Hydrogenation	Bis(benzonitrile)-dichloro palladium (II)	Amberlyst	118
Hydrogenation	$RhCl_3 \cdot 3H_2O$	Amberlyst A 21	118
Double bond shift	$RhCl_3 \cdot 3H_2O$	Amberlyst A 21	118
Transestrification	K_2PdCl_4	Amberlyst A 21	118
Vinylacetate synthesis	K_2PdCl_4	Amberlyst A 21	118
Hydroformylation	Tristiphenyl-Rhodium (I) Chloride	Chloromethylated polystyr-ene-resin reacted with chloro-diphenylphosphine	118
Hydroformylation	Hexachlororhodate(III)	Jonac XaX-1393	118
Hydroformylation	$RhCl_3 \cdot 3H_2O$	Amberlyst A 21	118
Hydroformylation	Rh or Ru-chloride	Poly-(p-diphenylphosphino) styrene	120
Oligomerization of olefins	Ni or Co-Salts	Poly(4-vinylpyridin)	119

An interesting application of organic polymers is their use as *catalyst carriers*, by which special properties can sometimes be transferred to the catalyst. The hydrogenation of several unsaturated compounds on palladium/polymer catalysts was measured, and a comparative study showed that the activity of the catalyst decreased in the following sequence as a function of the nature of the polymer [122]:

polyvinylalcohol > polyacrylonitrile > polyamid > polyester

Asymmetric hydrogenations have been reported with palladium on silk [123], palladium on modified cellulose [124] and on modified ion exchange resins [125]. Also with Raney Nickel modified with amino acids [126] and peptides [127]. Platinum-carbon catalysts exhibiting shape selectivity have been made by coating them with a thermosetting resin, which is carbonized. In such a way an organic molecular sieve skin is formed over the original catalyst [128].

Copper on a synthetic polypeptide demonstrated a remarkable selectivity in alcohol dehydrogenation by virtually excluding alcohols of complex structure such as diisopropyl and diisobutyl carbinol, while admitting simple alcohols such as n-butyl, isobutyl and sec-butyl [129].

Platinum on nylon appeared to be quite selective for the formation of cyclohexene on hydrogenation of benzene. On comparing different nylons, it was concluded that the spacings of the amide groups, which were thought to complex with the platinum atoms, were responsible for this selectivity [130].

Charcoal appeared to enhance the catalytic activity of supported Cobalt phthalocyanine for the oxidation of mercaptans [131]. The electron donor-acceptor complexes, which were shown to be such good catalysts for the activation of molecular hydrogen, were shown to be active also for other reactions like the hydrogenation of unsaturated hydrocarbons, the isomerization of butenes, the formation of ammonia from nitrogen and hydrogen and the formation of hydrocarbons from hydrogen and carbon monoxide. It was shown that graphite also acts as a good electron acceptor [132].

Polymeric enzymes and enzyme analogs constitute a completely separate topic. This field has recently been reviewed [133]. Some of the materials described act as heterogeneous catalysts and could also be considered as polymeric heterogeneous catalysts.

A typical Class C reaction, the valence isomerization of quadricyclane into norbornadiene, was shown to be catalyzed heterogeneously by metal complexes of phthalocyanine, tetraphenylporphyrin and several Schiff bases. This reaction is forbidden, because of a change in symmetry of the occupied orbitals during the reaction. It was shown that the ligand furnishes an orbital of the rigth symmetry in this case, by which a pathway for the electrons can be found. Those catalysts having the least energy separation between this orbital and the occupied metal orbitals, showed the highest catalytic activity [134].

11. Conclusion

From the discussion in the previous pages it has become clear that in spite of the many reactions that can be catalyzed by organic and metallo-organic catalysts heterogeneously, very few of these materials have found their way into practical application.

The study of these catalysts may sharpen our insight into the phenomenon of catalysis itself however. We have seen that the study of electron donor/acceptor complexes for the activation of molecular hydrogen has shown many correlations between physical properties of the acceptor and catalytic activity. In oxidation and dehydrogenation correlations were found with the redox properties of the solid. In the decomposition of nitrous oxide the paramagnetic properties, and with them, the catalytic activity, of organic polymers could be changed at will by modifications in the polymer structure, and in acid catalysis activity could be regulated by changing the type of acidic group and by selective neutralization.

It would have been difficult to find these kinds of relations with inorganic materials proper. The organic molecule is unique in that its physical properties may be changed in a continuous way by small modifications in its structure or by complexation with different metal ions. Another advantage is the fact that solid state effects are of smaller importance, and catalytic properties of the solid phase may be compared with physical properties in solution. In particular an extended π-electron system works as a catalytic entity in itself, irrespective of whether it is surrounded by other molecules of its kind (solid phase) or solvating molecules (solutions).

The great drawback of organic materials for practical use, especially at elevated temperatures, is their inherent instability, thermal as well as mechanical. On the other hand great advances have been made in polymer technology. Composite plastics are used in the construction of supersonic airplanes to replace metals, because of their superior resistance to heat, and plastic coatings are frequently used to enhance the chemical resistance of surfaces. Therefore it is not improbable that properly synthesized organic materials may also find their practical application as catalysts or catalyst carriers.

The work and ingenuity invested in the study of these materials may not only be of importance for a better understanding of the phenomenon of catalysis, but may also pave the way for the development of better and more selective catalysts.

12. References

[1] Hanke, W.: Heterogene Katalyse an halbleitenden organischen Verbindungen. Z. Chem. 9, 1 (1969).
[2] Roginski, S. Z., Sakharov, M. M.: Catalytic Properties of Organic Semiconductors. Russ. J. Phys. Chem. (English Transl.) 42, 696 (1968).

31

[3] Krause, H. W.: Organische Halbleiterkatalysatoren. Fortschr. Chem. Forsch. *6*, 327 (1966).
[4] Charentenay, F. X. de, Castel, P., Theyssié, Ph.: Polymères Semi-Conducteurs. Rev. Inst. Franc. Petrole Ann. Combust. *18*, 1126 (1963).
[5] Leftin, H. P., Hermana, E.: Optical Spectra of Adsorbed Molecules. Mechanism of Stereo Selective Olefin Isomerisation on Silica-Alumina. Proc. Third Intern. Congr. Catalysis. Vol. II, p. 1064. Amsterdam: North Holland Publ. Comp. 1965.
[6] Manassen, J., Klein, F. S.: Reactions of n-Butene and Butan-2-ol in Dilute Acid. The Elucidation of the Mechanism and the Intermediate in Elimination from Secondary Alcohols and in the Hydration of Olefins. J. Chem. Soc. *1960*, 4203.
[7] Szwarc, M.: Carban Ions, Living Polymers and Electron Transfer Processes. New York: Interscience Publ. 1968.
[8] Woodward, R. G., Hoffmann, R.: The Conservation of Orbital Symmetry. New York: Acad. Press 1969; Fukui, K.: Orientation and Stereoselection. Fortschr. Chem. Forsch. *15*, 1 (1970).
[9] Cassar, L., Eaton, Ph. E., Halpern, J.: Catalysis of Symmetry Restricted Reactions by Transition Metal Compounds. J. Am. Chem. Soc. *92*, 3515 (1970).
[10] Lugt, W. Th. A. M. van der: Symmetry Control and Transition Metal Catalyzed Reactions Tetrahedron Letters *1970*, 2281.
[11] Trapnell, B. M. W.: The Parahydrogen and Orthodeuterium Conversions and the Hydrogen-Deuterium Exchange. Catalysis, Vol. III (ed. Paul H. Emmett), p. 1. New York: Reinhold 1955.
[12] Cremer, E.: Heterogene Ortho- und Parawasserstoffkatalyse. Handb. Katalyse, Vol. VI (ed. G.-M. Schwab). Heterogene Katalyse III, p. 1. Wien: Springer Verlag 1943.
[13] Wigner, E.: Über die paramagnetische Umwandlung von Para-Orthowasserstoff III. Z. Phys. Chem. *B 23*, 28 (1933).
[14] Farkas, L., Sachsse, H.: Über die homogene Katalyse der Para-Orthowasserstoffumwandlung unter Einwirkung paramagnetischer Moleküle I, II. Z. Phys. Chem. *B 23*, 1, 19 (1933).
[15] Schwab, G.-M., Agallidis, E.: Über die Einwirkung von Organischen Radikalen auf Para-Wasserstoff. Z. Phys. Chem. *B 41*, 59 (1938).
[16] Bonhoeffer, K. F., Harteck, D.: Über Para- und Ortho-Wasserstoff. Z. Phys. Chem. *B 4*, 113 (1929).
[17] Rummel, K. W.: Über die Parawasserstoff-Umwandlung an Kohle-Oberflächen bei tiefen Temperaturen. Z. Phys. Chem. *A 167*, 221 (1933).
[18] Bonhoeffer, K. F., Farkas, A., Rummel, K. W.: Über die heterogene Katalyse der Parawasserstoffumwandlung. Z. Phys. Chem. *B 21*, 225 (1933).
[19] Calvin, M., Cockbain, E. G., Polanyi, M.: Activation of Hydrogen by Phthalocyanine and Copper-Phthalocyanine. Part. I. Trans. Faraday Soc. *32*, 1436 (1936).
[20] — Eley, D. D., Polanyi, M.: Activation of Hydrogen by Phthalocyanine and Copper-Phthalocyanine. Part. II. Trans. Faraday Soc. *32*, 1443 (1936).
[21] Eley, D. D.: The Conversion of Parahydrogen by Porphyrin Compounds, Including Hemoglobin. Trans. Faraday Soc. *36*, 500 (1940).
[22] Turkevich, J., Selwood, P. W.: Solid Free Radical as Catalyst for Ortho-Para-Hydrogen Conversion. J. Am. Chem. Soc. *63*, 1077 (1941).
[23] Harrison, L. G., McDowell, E. A.: The Catalysis of the Para-Hydrogen Conversion by the Solid Free Radical α,α'-diphenyl-β-picryl Hydrazyl. Proc. Roy. Soc. London Ser. A, *220*, 77 (1953).
[24] Eley, D. D., Inokuchi, H.: Organic Solids and Heterogeneous Catalysis. Electron Transfer in α,α'-diphenyl-β-Pycryl-Hydrazyl. Z. Elektrochem. *63*, 29 (1959).
[25] Turkevich, J., Laroche, J.: Catalytic Activity of a Graded Set of Charcoals for the Hydrogen-Deuterium Equilibration and the Ortho-Para Hydrogen Conversion and Electron Spin Resonance. Z. Phys. Chem. N.F. *15*, 399 (1958).

26) Acres, G. J. K., Eley, D. D.: Activation of Hydrogen by Polycopperphthalocyanine. Trans. Faraday Soc. *60*, 1157 (1964).

27) Davydova, I. R., Kiperman, S. L., Slinkin, A. A., Dulov, A. A.: On the Catalytic Activity of Certain Synthetic Organic Polymers Bull. Acad. Sci. USSR (English Transl.) *1964*. 1502

28) Tamaru, Kenzi: Catalysis by Electron Donor-Acceptor Complexes. Advan. Catalysis *20*, 327 (1969).

29) Kondow, T., Inokuchi, H., Wakayama, N.: Ortho-Para Hydrogen Conversion and Hydrogen-Deuterium Exchange in the Presence of Tetracyanopyrene-Cesium Complex. J. Chem. Phys. *43*, 3766 (1965).

30) Inokuchi, H., Wakayama, N., Kondow, T., Mori, Y.: Activated Adsorption of Hydrogens on Aromatic-Alkalimetal Charge-Transfer Complexes. J. Chem. Phys. *46*, 837 (1967).

31) – Mori, Y., Wakayama, N.: Hydrogen Exchange between Deuterium and Charge-Transfer Complexes. J. Catalysis *8*, 288 (1967).

32) – Wakayama, N., Hirooka, T.: Effect of Chemisorption of Hydrogen on Electrical Conductivity of Perylene-Cesium Charge-Transfer Complexes. J. Catalysis *8*, 383 (1967); *8*, 383 (1967).

33) Ichikawa, M., Soma, M., Onishi, T., Tamaru, K.: Reactivity of Electron Donor-Acceptor Complexes. Part. 8. Exchange Reaction of Hydrogen between Various Phthalocyanine EDA Complexes and Acetylene or Molecular Hydrogen. Trans. Faraday Soc. *63*, 1215 (1967).

34) – Reactivity of Electron Donor-Acceptor Complexes. Part. 6. Hydrogen Exchange of Aromatic Cyano-Substituted Compounds. Trans. Faraday Soc. *63*, 997 (1967).

35) Wakayama, N., Inokuchi, H.: Catalytic Behavior of Organic Semiconductors. Hydrogen Exchange on Aromatic-Alkali-Metal Charge-Transfer Complexes. J. Catalysis *11*, 143 (1968).

36) – Mori, Y., Inokuchi, H.: Catalytic Behavior of Organic Semiconductros. Relation Between Reactivity and Valency of Aromatic in Aromatic-Alkali Metal Complexes. J. Catalysis *12*, 15 (1968).

37) – Inokuchi, H.: Catalytic Behavior of Organic Semiconductors. Hydrogen Exchange on Barium-Naphthacene Ion Salts. J. Catalysis *15*, 417 (1969).

38) Ichikawa, M., Soma, M., Onishi, T., Tamaru, K.: Exchange Reactions of Hydrogen over the Electron Donor-Acceptor Complexes of Various Phthalocyanines with Sodium. J. Catalysis *6*, 336 (1966).

39) – Hydrogen Exchange Reaction between Molecular Hydrogen and the Electron Donor-Acceptor Complexes of Various Aromatic Compounds. Bull. Chem. Soc. Japan *40*, 1015 (1967).

40) Tanaka, S., Ichikawa, M., Naito, S., Soma, M., Onishi, T., Tamaru, K.: The Behavior of Hydrogen Chemisorbed over the Electron Donor-Acceptor Complexes of Aromatic Hydrocarbons with Sodium in the Hydrogen Exchange and Hydrogenation Reaction. Bull. Chem. Soc. Japan *41*, 1278 (1968).

41) Ichikawa, M., Soma, M., Onishi, T., Tamaru, K.: Reactivity of Electron Donor-Acceptor Complexes of Aromatic Compounds. Bull. Chem. Soc. Japan *40*, 1294 (1967).

42) – Reactivity of Electron Donor-Acceptor Complexes. Part 9. Isomerisation of Butene. Trans. Faraday Soc. *63*, 2012 (1967).

43) – Reaction between Molecular Hydrogen and Various Electron Donor-Acceptor Complexes of Aromatic Hydrocarbons with Sodium. Their Electronic Properties. J. Am. Chem. Soc. *91*, 6505 (1969).

44) Haber, F., Weiss, J.: The Catalytic Decomposition of Hydrogen Peroxide by Iron Salts. Proc. Roy. Soc. (London), Ser. A, *147*, 332 (1934).

45) Gallard-Nechtstein, J., Salle, R., Traynard, Ph.: Catalyse sur Polychélates. C. R. Acad. Sci. Paris, Ser. C, *262*, 949 (1966).

46) Cook, A. H.: Catalytic Properties of the Phthalocyanines. Part. I. Catalase Properties. J. Chem. Soc. *1938*, 1761.
47) Roginskii, S. Z., Berlin, A. A., Golovina, O. A., Dokukina, E. S., Sakharov, M. M., Cherkashina, L. G.: Catalytic Activity of Polyphthalocyanines with Respect to the Decomposition of Hydrogen Peroxide. Kinetica Katal. *4*, 431 (1963); C. A. *59*, 5833d (1963).
48) Levina, S. D., Andrianova, T. I., Sakharov, M. M., Golovina, O. A., Lobanova, K. P., Rotenberg, Z. A.: Catalytic Properties of Systems Consisting of Powered Metals and Phthalocyanines. Russ. J. Phys. Chem. (English Transl.) *40*, 660 (1960).
49) Keier, N. P., Troitskaya, M. G., Rukhadze, E. G.: Catalytic Activity of Chelate Polymers in the Reaction of Hydrogen Peroxide Decomposition. Kinetika Katal. *3*, 691 (1962); C. A. *58*, 7413a (1963).
50) Hercules, David M.: Electron Spectroscopy. Anal. Chem. *42*, 20 A (1970).
51) Topchiev, A. V., Geiderikh, M. A., Davydov, B. E., Kargin, V. A., Krentsel, B. A., Kustanovich, I. M., Polak, L. S.: Preparation of Polymeric Materials with Semiconductor Properties Based on Acrylonitrile. Dokl. Akad. Nauk SSSR *128*, 312 (1959); C. A. *54*, 6223h (1960).
52) Berlin, A. A., Blyumenfeld, L. A., Semenov, N. N.: Catalytic Properties of some Macromolecular Structures. Izv. Akad. Nauk SSSR, Ser. Khim. *1959*, 1689. C. A. *54*, 8252h (1960).
53) Inoue, H., Kida, Y., Imoto, E.: Vulcanized Anilinblack. Its Electrical Conductivity and Catalysis upon the Decomposition of Hydrogen Peroxide. Bull. Chem. Soc. Japan *39*, 551 (1966).
54) Parini, V. P., Kazokova, Z. S., Berlin, A. A.: Polymers with Conjugated Bonds and Heteroatoms in the Conjugated Bond Chain. XIX. Some Properties of Aniline Black. Vysokomolekul Soedin. *3*, 1870 (1961). C. A. *56*, 14460g (1962).
55) Dokukina, E. S., Golovina, O. A., Sakharov, M. M., Aseeva, R. M.: Catalytic Activity of Organic Semiconductors Obtained by Thermal Dehydrochlorination of Poly(vinylchloride). Kinetika Katal. *7*, 660 (1966). C. A. *65*, 17755f (1966).
56) Manassen, J., Wallach, J.: Organic Polymers. Correlation between their Structure and Catalytic Activity in Heterogeneous Systems. I. Pyrolyzed Polyacrylonitrile and Polycyanoacetylene. J. Am. Chem. Soc. *87*, 2671 (1965).
57) Higginson, W. C. E.: The Oxidation of Hydrazine in Aqueous Solution. The Chemical Society, London. Special Publication No. 10, p. 95 (1957). (Recent Aspects of the Inorganic Chemistry of Nitrogen.)
58) Willhoft, E. M. A., Robertson, A. J. B.: Mass-Spectrometric Investigation of Di-imide by the Catalytic Decomposition of Hydrazine at Low Pressures on Platinum. Chem. Commun. *1967*, 385.
59) Boreskov, G. K., Keier, N. P., Rubtsova, L. P., Rukhadze, E. G.: The Catalytic Properties of Chelate (Intramolecular) Polymers. Dokl. SSSR. Phys. Chem. Section *144*, 453 (1962). (English Transl.)
60) Keier, N. P., Boreskov, G. K., Rubtsova, L. F., Rukhadze, E. G.: Catalytic Activity of Organic Polymers. III. Regularities of Catalysis on the Chelate Polymers of Different Chemical Composition and Structure. Kinetika Katal. *3*, 680 (1962). C. A. *58*, 7412g (1963).
61) – Regularities of Catalysis on Chelate Polymers. Proc. Third Intern. Congr. Catalysis, p. 1021. Amsterdam: North Holland Publishing Co.
62) – Astafiev, I. V.: Catalytic Activity of Organic Polymers. II. Catalytic Proporties of a Polymer Obtained by Dechlorination of Poly(vinylidene Chloride). Kinetika Katal. *3*, 364 (1962).
63) Hine, J.: Physical Organic Chemistry (Second Edition), p. 311. New York: McGraw-Hill Book Company Inc. 1962.

64) Sachtler, W. M. H., Fahrenfort, J.: The Catalytic Decomposition of Formic Acid Vapor on Metals. Proc. Second. Intern. Congr. Catalysis. Technip. Paris 1961, p. 831.

65) Walling, Ch.: Free Radicals in Solution, p. 273 and 288. New York: John Wiley and Sons Inc. 1957.

66) Beckwith, A. L. S., Norman, R. O. C.: Electron Spin Resonance Studies. Part XX. The Generation of Organic Radicals by the One Electron Reduction of Aliphatic Halogeno-Compounds in Aqueous Solution. J. Chem. Soc. A 1969, 400.

67) Hanke, W.: Katalyse an Phthalocyaninen. I. Ameisensäuredampfzerfall an Metall-Phthalocyaninen. Z. Anorg. Allgem. Chem. 343, 121 (1966).

68) – Katalyse an Phthalocyaninen. II. Ameisensäuredampfzerfall an polymeren Kupfer-Phthalocyanin und die Bedeutung des elektronischen Faktors. Z. Anorg. Allgem. Chem. 347, 67 (1967).

69) – Katalyse an Phthalocyaninen IV. Mangan-Formiat-Phthalocyanin und der Ameisensäuredampfzerfall an Mangan Phthalocyanin. Z. Anorg. Allgem. Chem. 355, 160 (1967).

70) – Gutschick, D.: Katalyse und Phthalocyaninen V. Über ein Zinkphthalocyanin-Ameisensäure-Addukt und zur Wechselwirkung von Metallphthalocyaninen mit Ameisensäure. Z. Anorg. Allgem. Chem. 366, 201 (1969).

71) Rosswurm, H., Haevecker, R., Doiwa, A.: Katalyse an organischen Halbleitern I. Ameisensäuredampfzerfall an Phthalocyaninen. Z. Anorg. Allgem. Chem. 350, 1 (1967).

72) Tanaka, S., Onishi, T., Tamaru, K.: Decomposition of Formic Acid over Metal Phtahlocyanines. Bull. Chem. Soc. Japan 41, 2557 (1968).

73) Roginskii, S. Z., Berlin, A. A., Kutseva, L. N., Aseeva, R. M., Cherkashina, L. G., Sherle, A. I., Matseeva, N. G.: Catalytic Properties of Organic Polymers with a Conjugated Bond System. The Formation of Hydroperoxides by Oxidation of Alkylaromatic Hydrocarbons and Cychlohexane. Dokl. SSSR, Chemistry Section (English Transl.) 148,35 (1963) (1963).

74) Nesmeyanov, A. N., Rubinstein, A. M., Dulov, A. A., Slinkin, A. A., Rubinskaya, M. I., Slonimskii, G. L.: Catalytic Properties of Polymers Prepared from Methyl-Chlorovinyl Ketone. Dokl. Akad. Nauk. SSSR 135, 609 (1960). C. A. 55, 15338a (1961).

75) Taylor, William F.: Catalysis in Liquid-Phase Autoxidation. I. Effect of Polymeric Surfaces. J. Catalysis 16, 20 (1970).

76) Hock, H., Kropf, H.: Autoxidation von Kohlenwasserstoffen XXVII. Phthalocyanine als Katalysatoren für die Autoxydation von Cumol und sonstigen Benzolkohlenwasserstoffen. J. Prakt. Chem. 9, 173 (1959).

77) Kropf, H.: Katalyse durch Phthalocyanine. I. Kinetik und Mechanismus der Autoxydation von Cumol in Gegenwart von Phthalocyaninen. Ann. Chem. 637, 73 (1960).

78) – Katalyse durch Phthalocyanine. II. Katalyse von Benzolkohlenwasserstoffen durch Kupfer-Phthalocyanine. Ann. Chem. 637, 93 (1960).

79) – Katalyse durch Phthalocyanine. V. Autoxydation von Cumol in Gegenwart von Vanadyl-Phthalocyanin. Tetrahedron Letters 1962, 577.

80) – Hoffmann, Hd.: Autoxydation von Cumol in Gegenwart von substituierten Kupfer-Phthalocyaninen und verwandten Kupfer-Komplexen. Tetrahedron Letters 1967, 659.

81) Minkov, A. I., Keier, N. P.: The Mechanism of Cumene Oxidation over Copper Phthalocyanine. Kinetika Katal. 8, 160 (1967). C. A. 67, 21210s (1967).

82) – – Anufrienko, V. F.: Oxidation of Isopropylbenzene on Copper Phthalocyanine. Kinetika Katal. 8, 387 (1967). C. A. 67, 107899 (1967).

83) Kundo, N.N., Keier, N. P.: Catalytic Activity of Organic Copper Complexes in Cysteine Oxidation. Kinetika Katal. 8, 796 (1967). C. A. 68, 30008e (1968).

84) Wolberg, A., Manassen, J.: Electrochemical and Electron Paramagnetic Resonance Studies on Metalloporphyrins and their Electrochemical Oxidation Products. J. Am. Chem. Soc. 92, 2982 (1970).

85) Falk, J. E.: Porphyrins and Metalloporphyrins. p. 26. Amsterdam: Elsevier 1964.
86) Kamiya, Y.: The Autoxidation of α-Methylstyrene, Catalyzed by Copperphthalocyanine. Tetrahedron Letters *1968*, 4965
87) Inoue, H., Kida, Y., Imoto, E.: The Catalytic Action of Binary Metal-Polyphthalocyanine Complexes on the Oxidation of Acetaldehyde Ethylene Acetal. Bull. Chem. Soc. Japan *38*, 2214 (1965).
88) − − − Organic Catalysts. III. The Catalytic Action of Copper-Iron Polyphthalocyanine on Oxidation of Acetalhyde Ethylene Acetal. Bull. Chem. Soc. Japan *40*, 184 (1967).
89) − − − Organic Catalysts. The Role of the Iron as an Oxidation Catalysts in Copper-Iron Polyphthalocyanine. Bull. Chem. Soc. Japan *41*, 684 (1968).
90) − − − Organic Catalysts. V. Specific Catalytic Properties of Copper-Iron Polyphthalocyanine in the Oxidation of Aldehydes. Bull. Chem. Soc. Japan *41*, 692 (1968); *41*, 692 (1968).
91) Keier, N. P., Mamaeva, E. K., Alikina, G. M., Tyuleneva, L. I., Afanaseva, S. M.: Catalytic Activity of the Transition Metal Chelate Salts of Quinaldine Dithioamides on Oxidation of Cumene. Kinetika Katal. *6*, 849 (1965).
92) Minkov, A. I., Alikina, F. M., Fridnev, Yu. M., Keier, N. P.: Catalytic Activity of Polychelates in the Liquid-Phase Oxidation for Hydrocarbons. Kinetika Katal. *7*, 632 (1966).
93) Universal Oil Products Co.: Oxidative Sweetening of Kerosine with Phthalocyanine Catalysts. U.S. Patent 3, 408, 287 (1968); C.A. *70*, 13245d (1969).
94) Allis Chambers Manufg. Co.: Fuel Electrodes Having a Metal Phthalocyanine Catalyst. U.S. Patent 3, 410, 727 (1968); C.A. *70*, 33823u (1969).
95) Compagnie Francaise de Raffinage (1966): Catalyseurs Organiques a Grande Spécifique et leur Procédé de Fabrication. French Patent 1, 431, 848.
96) Manassen, J., Khalif, Sh.: Organic Polymers. Correlation between their Structure and Catalytic Activity in Heterogeneous Systems. II. Models for Dehydrogenation Catalysts. J. Am. Chem. Soc. *88*, 1943 (1966).
97) − Catalytic Hydrogen Transfer. Israel Patent Apllication 27693 (1966).
98) Berlin, A. A., Ragimov, A. V., Liogonkii, B. I., Belova, G. V.: Synthesis and Study of Polyarylene Quinones. Polymer Sci. (USSR) (English Transl.) *8*, 589 (1966).
99) Manassen, J., Khalif, Sh.: Organic Polymers: Correlation between their Structure and Catalytic Activity in Heterogeneous Systems. IV. Oxydative Dehydrogenation. A Comparison between the Catalytic Activity of an Organic Polymer and that of some Molybdate Catalysts. J. Catalysis *13*, 290 (1969).
100) − Bar Ilan, A.: Metal Complexes of Phthalocyanine and α, β, γ, δ-Tetraphenyl Prophyrin as Heterogeneous Catalysts in Oxidative Dehydrogenation. Correlation between Catalytic Activity and Redox Potential. J. Catalysis *17*, 86 (1970).
101) Bar Ilan, A., Manassen, J.: Unpublished Results.
102) Jolly, W. L.: The Inorganic Chemistry of Nitrogen, p. 71. New York, Amsterdam: W. A. Benjamin Inc. 1964.
103) Mme Gallard, J., Laederlich, Th., Salle, R., Traynard, Ph.: Polymeres a Structures Conjuguée. I. Catalyse par les Polymères Conjugées. Bull. Soc. Chim. France *1963*, 2204.
104) − Nechtstein, M., Soutif, M., Traynard, Ph.: Polymères à Structure Conjuguée. II. Conclusion entre Propriétés Cataliques et Centres Paramagnétiques dans les Polymères Conjuguées. Bull. Soc. Chim. France *1963*, 2209.
105) Gallard-Nechtstein, J., Pècher-Reboul, A., Traynard, Ph.: Polymères à Structure Conjuguée. III. Influence de la Structure sur les Propriétes Cataliques des Polymères Conjugués. Bull. Soc. Chim. France *1967*, 960
106) − − − Heterogeneous Catalysis on Organic Conjugated Polymers. II. Electron Spin Resonance and Structural Factor. J. Catalysis *13*, 261 (1969).
107) Cutlip, M. B., Peters, M. S.: Heterogeneous Catalysis over an Organic Semiconducting

Polymer made from Acrylonitrile. Symp. on Recent Advances in Reaction Kinetics and Catalysis, Part. I. Sixtieth Annual Meeting, New York. November 26–30, 1967.

108) Paushkin, Ya. M., Burova, L. M., Voronina, M. A., Vishnyakova, T. D., Sokolinskaya, T. A., Aliev, L. A.: Dehydration and Dehydrogenation of Alcohols, Catalyzed by Ferrocene Polymers. Dokl. Akad. Nauk. SSSR *186*, 108 (1969).

109) Fang, F. T.: Heterogeneous Catalysis by Macroreticular Functional Resins. Proc. Third. Internat. Cong. Catalysis, p. 901. Amsterdam: North Holland Publ. Comp. 1965.

110) Manassen, J., Khalif, Sh.: Organic Polymers. Correlation between their Structure and Catalytic Activity in Heterogeneous Systems. III. Acid Type Catalysis, Sulfonated and Phosphonated Polyphenyl. J. Catalysis *7*, 110 (1967).

111) Kallo, D., Preszler, I.: n-Butene Isomerization on Acidic Ion-Exchange Resins. J. Catalysis *12*, 1 (1968).

112) Gates, B. C., Schwab, G.-M.: The Dehydration of Formic Acid Catalyzed by Polystyrene Sulfonic Acid. J. Catalysis *15*, 430 (1969).

113) – Johanson, L. N.: The Dehydration of Methanol and Ethanol Catalyzed by Polystyrene Sulfonate Resins. J. Catalysis *14*, 69 (1969).

114) Frilette, V. J., Mower, E. M., Rubin, M. K.: Kinetics of Dehydration of Ter-butyl Alcohol Catalyzed by Ion Exchange Resins. J. Catalysis *3*, 25 (1964).

115) Cassidy, H. G., Kun, K. A.: Oxidation-reduction Polymers; Redox Polymers. Interscience, New York (1965).

116) Chelate Polymers Give New Redox Route. Chem. Eng. News, April 2, 1962, p. 48.

117) Allan, G. G., Neogi, A. N.: Macromolecular Organometallic Catalysis. J. Phys. Chem. *73*, 2093 (1969).

118) Mobil Oil: Procédé de Mise en Oeuvre de Reactions Catalysées par und Complèxe Résine-Metal' Belgian Patent 721, 696 (1969).

119) I. C. I.: Verfahren zur Oligomerisation von Olefinen. German Patent 1,937,232 (1970).

120) Manassen, J.: Heterogeneous Hydroformylation of Olefins. Israel Patent Application 30505 (1968).

121) Legzdins, P., Rempel, G. L., Wilkinson, G.: The Protonation of Metal Carboxylates. New Homogeneous Hydrogenation Catalyst. Chem. Commun. *1969*, 825.

122) Tyruenkova, O. A.: Hydrogenation of Unsaturated Compounds on Palladium/Polymer Catalysts. Russian J. Phys. Chem. (English Transl.) *43*, 1167 (1969).

123) Akabori, S., Sakurai, S., Izumi, Y., Fujii, Y.: Asymmetric Catalyst. Nature *178*, 323 (1956).

124) Harada, K., Yoshida, T.: Asymmetric Hydrogenation Using Modified Cellulose-Palladium Catalysts. Naturwissenschaften *57*, 131 (1970).

125) – – Asymmetric Hydrogenation Using Modified Ion Exchange Resin-Palladium Catalysts. Naturwissenschaften *57*, 306 (1970).

126) Izumi, Y., Akabori, S., Fukawa, H., Tatumi, S., Imaida, N., Fukuda, T., Komatu, S.: Asymmetric Hydrogenation with Modified Raney Nickel. Proc. Third Intern. Congr. Catalysis. Volume II, p. 1364. Amsterdam: North Holland Publ. Comp. 1965.

127) – Tatsumi, S., Imaida, M., Okubo, K.: Asymmetric Hydrogenation of C=O Double Bonds with Modified Raney Nickel. XIII. Modification with Peptides. Bull. Chem. Soc. Japan *43*, 556 (1970).

128) Cooper, B. J.: Platinum-Carbon Catalysts with Molecular Sieve Properties. Shape Selectivity in Hydrogenation Catalysis. Platinum Metals Rev. *14*, 133 (1970).

129) Welch, R. C. W., Rase, H. F.: Selectivity Characteristics of a Geometrically Designed Heterogeneous Catalyst. A High Melting Copper-Enzyme Model. Ind. Eng. Chem. Fundamentals *8*, 389 (1969).

130) Harrison, D. D., Rase, H. F.: Nylon Platinum Catalysts with Unusual Geometric and Selective Characteristics. Ind. Eng. Chem. Fundamentals *6*, 161 (1967).

J. Manassen

131) Mme Nechtstein, J.: L'Oxidation du n-butyl-sulfure Catalysée par un mélange de Charbon Actif et de Phthalocyanine de Cobalt. C. R. Acad. Sci. Paris, Ser. C. *268*, 376 (1969).
132) Ichikawa, M., Sudo, M., Soma, M., Onichi, T., Tamaru, K.: Catalytic Formation of Hydrocarbons (C_1-C_5) from Hydrogen and Carbon Monoxide over the Electron Donor-Acceptor Complex Films of Alkalimetals with Transition Metal Phthalocyanines or Graphite. J. Am. Chem. Soc. *91*, 1538 (1969).
133) Lindsey, Alan S.: Polymeric Enzymes and Enzyme Analogs. J. Makromol. Chem. *C 3*, 1 (1969).
134) Manassen, J.: Catalysis of a Symmetry Restricted Reaction by Transition Metal Complexes. The Importance of the Ligand. J. Catalysis *18*, 38 (1970).

Received January 7, 1971

Catalytic Olefin Disproportionation

Dr. Robert L. Banks

Phillips Petroleum Company, Research and Development Department
Bartlesville, Oklahoma, USA

Contents

I. Introduction

Banks and Bailey [1] reported in 1964 that linear olefins could be catalytically converted in a *highly specific and efficient* manner to approximately equal molar quantities of shorter and longer chain olefins. For example, they found that over a cobalt-molybdate catalyst, propylene was disproportionated to a product containing 49 mole per cent ethylene, 48 per cent n-butenes, and 3 per cent C_5+ material. This novel reaction is general for hydrocarbons containing ethylenic carbon-carbon bonds which has been confirmed by continued investigation. An abundance of data and mechanistic studies by several groups of investigators have indicated the reaction proceeds by two unsaturated pairs of carbon atoms combining in a four-center transition state which dissociates by breaking the opposite set of bonds to form the new olefins.

The initial catalysts disclosed for olefin disproportionation were of the heterogeneous type: molybdenum and tungsten hexacarbonyls and oxides supported on alumina [1-3]. Homogeneous catalyst systems active for disproportionating internal olefins were reported in 1967 by Calderon and coworkers [4] and for disproportionating terminal and internal olefins were reported in 1968 by Zuech [5]. A number of catalysts are now known to be active for disproportionating olefins. A noncatalytic counterpart to the disproportionation reaction was reported in 1931 by Schneider and Frölich [6]; in the pyrolysis of propylene at 852 °C extrapolation of product distribution to zero conversion indicated the combination of two molecules of propylene to give one of ethylene and one of butene was a primary reaction that accounted for 48 per cent of the reacting propylene.

The reaction described by Banks and Bailey was of the type

$$2\,C_nH_{2n} \ \rightleftharpoons \ C_{(n-a)}H_{2(n-a)} \ + \ C_{(n+a)}H_{2(n+a)}$$

where $n \geqslant 3$ and $(n-a) \geqslant 2$, and was referred to by the authors as *"olefin disproportionation"*. The reaction has also been referred to as the "olefin reaction" [7]. Others use different terminology in subsequent publications: Bradshaw and coworkers [8] prefer "olefin dismutation"; they also use the term "ethenolysis" for the reverse reaction when one of the reactants is ethylene; Calderon and associates [4] use "olefin metathesis"; and Crain [9] uses "mutual cleavage" and "ethylene cleavage". As noted in a review by Bailey [10], the term "olefin disproportionation" has been widely used in the literature for these and related type reactions of unsaturated hydrocarbons that apparently proceed by the four-center mechanism.

II. Heterogeneous Disproportionation Catalysts

The catalysts reported by Banks and Bailey [1] consisted of a high surface area support, *alumina, on which molybdenum or tungsten compounds* had been deposited. Activation of the catalyst, or the support, was accomplished by heating at elevated temperature in a stream of air or nitrogen. A number of heterogeneous catalysts of this type are now known to be active for disproportionating olefins. Examples of the disproportionation of propylene to ethylene and *n*-butene with some of these catalysts are presented in Table 1. It should be noted that these examples were obtained by several investigators and over wide ranges of process conditions; thus quantitative comparisons of the various catalysts based on these data may not be justified.

Table 1. *Heterogeneous olefin disproportionation catalysts*

Catalyst	Test Conditions				Propylene Conv., %	Select. %	Ref. [b]
	Temp. °C	PSIG	Space Rate [a]				
$Mo(CO)_6 \cdot Al_2O_3$	121	500	500	G	25	97	1, 2)
$MoO_3 \cdot Al_2O_3$	50		1600	G	11.0	100	13/1, 3, 11)
$CoO \cdot MoO_3 \cdot Al_2O_3$	163	450	8.5	W	42.9	94.1	1, 12/14)
$MoO_3 \cdot AlPO_4$	538	0	2	W	5 >	95	12)
$MoO_3 \cdot SiO_2$	538	0	3.5	W	28 >	95	12, 15/16)
$MoO_3 \cdot Al_2O_3 \cdot TiO_2$	121	0	2	W	15		12)
$MoO_3 \cdot MgO \cdot TiO_2$	177	0	2	W	3		12/17)
$MoO_3 \cdot Mg_3Si_4O_{11} \cdot H_2O$	177	0	2	W	3		12)
$MoS_2 \cdot Al_2O_3$	149	0	600	G	1.3	100	12,18)
$MoS_2 \cdot SiO_2$	538	0	600	G	9.1	100	12,18)
$MoO_3 \cdot CrO_3 \cdot Al_2O_3$	160	90	180	G	36.0	97	19)
$WO_3 \cdot Al_2O_3$	177	0	600	G	7.4	100	3)
$WO_3 \cdot AlPO_4$	538	100	7.5	W	34	82	12)
$WO_3 \cdot SiO_2$	426	450	40	W	44.8	97.8	12, 15, 16)
$WO_3 \cdot SiO_2 \cdot Al_2O_3$	260	450	7200	G	43.5		15)
$WO_3 \cdot Al_2O_3 \cdot ThO_2$	200	0	2	W	1		12)
$WO_3 \cdot ThO_2$	200	0	2	W	3		12/20)
$WO_3 \cdot ZrO_2$	538	0	2	W	2		12)
$WO_3 \cdot Zr_3(PO_4)_4$	427	0	2	W	4		12)
$WO_3 \cdot Mg_3(PO_4)_2$	538	0	2	W	3		12)
$WO_3 \cdot Ti_3(PO_4)_4$	538	0	2	W	1		12)
$WO_3 \cdot Ca_3(PO_4)_2$	538	0	2	W	3		12)
$WS_2 \cdot Al_2O_3$	149	0	1	W	1.0	100	12, 18)
$WS_2 \cdot SiO_2$	538	0	600	G	18.3	100	12/18)

Table 1 (continued)

Catalyst	Test Conditions				Propylene Conv., %	Select. %	Ref.[b]
	Temp. °C	PSIG	Space Rate [a]				
$Re_2(CO)_{10} \cdot Al_2O_3$	100		1600	G	20.4	100	21)
$Re_2O_7 \cdot Al_2O_3$	25		200	G	19.2	100	11,22,23,24)
$Re_2O_7 \cdot SiO_2$	204	0	2	W	4.0	100	12/25)
$Re_2O_7 \cdot ZrO_2$	150		1600	G	2.1	100	25)
$Re_2O_7 \cdot ThO_2$	150		1600	G	12.0	100	25)
$Re_2O_7 \cdot SnO_2$	50		1600	G	15.0	100	25/26)
$Re_2O_7 \cdot TiO_2$	150		1600	G	13.0	100	25)
$Re_2O_7 \cdot Fe_2O_3$	250		1600	G	0.4	100	27)
$Re_2O_7 \cdot NiO$	150		1600	G	3.9	100	27)
$Re_2O_7 \cdot WO_3$	250		1600	G	0.75	100	27)
$SnO_2 \cdot Al_2O_3$	350		1600	G	1.5		28)
$La_2O_3 \cdot Al_2O_3$	350		1600	G	1.9		28)
$Rh_2O_3 \cdot Al_2O_3$	350		1600	G	6.0		28)
$Os_2O_3 \cdot Al_2O_3$	350		1600	G	0.5		28)
$Ir_2O_3 \cdot Al_2O_3$	350		1600	G	3.5		28)
$V_2O_3 \cdot SiO_2$	560		1000	G	11.2	44	29)
$Nb_2O_5 \cdot SiO_2$	538	0	600	G	3.7	90	29/12)
$Ta_2O_5 \cdot SiO_2$	538	0	600	G	8.3	56	29/12)
$TeO_3 \cdot SiO_2$	538	450	20	W	20		12)
Al_2O_3	315		0.3	W	0.9	100	30/31)
MgO	435	300	30	W	1.5	100	32)

[a] G – Gas hourly space velocity.

W – Weight hourly space velocity.

[b] Numbers after slant mark are references to additional examples.

Studies by Lapidus and coworkers [33] of the isomerization of *n*-butenes on nickel-zeolite catalysts indicate that some *zeolite catalysts* are active for the disproportionation of butenes to propylene and pentenes.

Composition and Physical Properties

Surface areas reported for heterogeneous disproportionation catalysts were generally greater than 100 m²/g and the promoter concentrations reported were generally in the range of 1 to 15 per cent. Typical compositions and physical properties of three disproportionation catalysts are shown in Table 2.

Table 2. *Compositions and physical properties of typical disproportionation catalysts*

	$CoO \cdot MoO_3 \cdot Al_2O_3$	$Re_2O_7 \cdot Al_2O_3$	$WO_3 \cdot SiO_2$
Composition, Wt. %			
MoO_3	11.0		
Re_2O_7		14	
WO_3			6.8
CoO	3.4		
Al_2O_3	85.6	86	
SiO_2			93.2
Surface area, m^2/g	284	255[a]	345
Pore volume, cm^3/g	0.58		0.98
Avg. pore diameter, Å	82	58[a]	114
Ref.	1)	11)	15)

[a] Of the support before acid treatment.

Catalyst Preparation

Relatively little has been reported on optimization of catalyst preparation procedures. Catalysts prepared by conventional impregnation methods have been described [1, 12, 16]. Supported molybdenum and tungsten hexacarbonyl catalysts were prepared by impregnating at 65 °C under vacuum preactivated gamma-alumina with cyclohexane solution of the hexacarbonyls and removing the cyclohexane from the catalyst by flushing with nitrogen and treating under vacuum at 120 to 150 °C [2]. Preactivation of the alumina was at 538 °C with air followed by nitrogen. Rhenium catalysts have been prepared by impregnating alumina with a solution of rhenium heptaoxide in an inert organic solvent [34], decomposing ammonium perrhenate to rhenium heptaoxide and subliming the heptaoxide onto preactivated alumina [13, 23], fluidizing at 150 °C rhenium carbonyl with preactivated alumina [21], and ball milling rhenium oxide with silica [12]. The ball mill technique has also been used to prepare supported sulfide catalysts [12]. Patents issued to British Petroleum Company include examples showing that rhenium oxide-alumina catalysts prepared by acid-treating the alumina support, by using freshly prepared alumina, or by subliming rhenium oxide onto alumina, were more active for disproportionating propylene than catalysts prepared by conventional impregnation of commercial calcined alumina (Table 3).

Table 3. *Preparation of rhenium oxide-alumina catalysts*

Alumina	14 % Re_2O_7 added by	Propylene conv., % [a]	Ref.
Commercial (Gamma)	Impregnation	19.2	11, 23, 24)
Acid-treated (0.1 N HCl)[b]	Impregnation	31.0	11)
Freshly precipitated	Impregnation	28.8	24)
Commercial (Gamma)	Sublimation	34.7	23)

[a] Test at 25 °C and 2000 GHSV.
[b] Similar results with acetic acid treatment.

Activation and Regeneration

Heterogeneous disproportionation catalysts are generally activated at elevated temperature in a stream of dry air or inert gas. This procedure removes water and absorbed compounds that may inhibit the disproportionation reaction, and possibly promotes the formation of specific atomic configurations on the surface. Typically activation is carried out at 500 to 600 °C for two to ten hours. However, wide ranges of temperatures and times have been used and have been shown to affect activity: the elevation of activation temperature from 580 to 800 °C increased propylene disproportionation over a rhenium oxide-stannic oxide catalyst from 19.0 per cent to 29.6 per cent [26]. Catalysts deactivated during disproportionation were reactivated or regenerated by a repetition of the activation procedure, using a controlled amount of oxygen to burn-off the accumulated coke [3, 15]. One tungsten oxide-silica catalyst after 110 regeneration cycles over about a one-year period was reported to be still active and selective for propylene disproportionation [15]. Examples with some catalysts show that treatment of the activated catalyst with reducing gases, such as carbon monoxide and hydrogen, increased activity and/or selectivity; however, such treatments were not necessary to obtain good activity [15, 35]. Lester [36] reports that a cobalt molybdate catalyst was more active for disproportionating propylene when cooled from activation to process temperature in the presence of an oxidizing agent (air) than when cooled by nitrogen purging.

Catalyst Poisons

Various polar and chemical compounds reportedly are capable of poisoning or deactivating disproportionation catalysts if present in the feed or allowed to contact the catalyst after activation. For example, propylene conversion over cobalt-molybdate catalyst was reduced when 300–2000 ppm of oxygen, water, carbon dioxide, hydrogen sulfide, ethyl sulfide, acetylene, or propadiene

was in the feed [3]. Tungsten oxide-silica which operates at a higher temperature, is reported to have considerable resistance to poisons; water, air, acetone,
carbon monoxide, hydrogen, or methanol in the feed reduced activity, but
original activity was restored on introduction of pure feed [16].

Catalyst Modifications

Increasing selectivity of disproportionation catalysts by incorporation of minor amounts of alkali or alkaline earth metal ions to reduce double-bond isomerization and other acid-type reactions have ben studied by several investigators. Bradshaw and coworkers [8, 37] used sodium bicarbonate solutions of
various strengths to poison the acid sites of cobalt molybdate catalyst; Crain
[9] treated molybdena-alumina catalyst with aqueous potassium hydroxide to
obtain highly selective catalyst for disproportionating 3-heptene, 1-octene,
and 2-octene, and for cleaving 2-octene and 2,3-dimethylbutene-2 with ethylene. Heckelsberg [38] used Na, K, Ba, and Cs with tungsten oxide-silica and
molybdena-silica to increase selectivity for disproportionating propylene.
Ray and Crain [58] and Mango [39] treated cobalt molybdate-alumina with
aqueous potassium hydroxide solution to improve selectivity for reacting
cycic olefins with acyclic monoolefins. Van Helden and coworkers [14] incorporated K, Na, Rb, and Cs with cobalt-molybdate and K with rhenium oxide-
alumina to increase selectivity to primary products.

Increased disproportionation activity of tungsten oxide-silica following
treatment of the catalyst with hydrogen chloride, or hydrocarbon chlorides
that decomposed to hydrochlorides at the temperature of the treatment, was
disclosed by Pennella [40]. Table 4 shows propylene conversions before and
after 20–80 minute treatments of the catalysts with various chloride compounds.

Table 4. *Propylene conversions before and after treatment of catalyst with various chloride compounds*

Tungsten oxide-silica treated at 500 °C with	% Propylene conversion Before treatment	After treatment
Hydrogen Chloride	10.4	44.8
Vinyl Chloride	2	42
Chloropropane	13.4	42.5
Chlorobenzene	12	21

Use of triethylaluminum to maintain the activity of cobalt molybdate catalyst for disproportionation of 1-butene is reported in a patent issued to Shell
International [41]. Tributylphosphine was used with tungsten oxide-silica cata-

lyst by Heckelsberg [42] to obtain higher conversions of 2-pentene. A patent issued to Alkema and Van Helden [43] relates to adding molecular hydrogen to the system to improve catalytic disproportionation of acyclic alkenes. Some of the examples indicate that large fractions of alkanes were produced.

Bifunctional Catalyst Systems

Reduction of double-bond isomerization activity resulted in high selectivity to primary disproportionation products; however, in some applications good double-bond isomerization activity is essential, e.g. process for producing detergent range olefins from propylene. Symmetrical olefins reacting with themselves do not yield new olefins; hence, a double-bond shift is needed prior to the disproportionation reaction. Incorporation of an acid-type double-bond catalyst is undesirable for some applications as it would also promote skeletal isomerization and polymerization. Magnesium oxide is a very selective catalyst for double-bond isomerization, and Banks and Kenton [44] obtained high disproportionation conversions by mechanically mixing magnesia with disproportionation catalyst. Results obtained with 2-pentene feed are shown in Table 5. Banks and Kenton also showed that magnesia could be used at ambient temperature to treat the olefin feed to obtain higher conversions[44].

Table 5. *Disproportionation of 2-pentene with and without selective double-bond isomerization catalyst*

Pressure, psig[a]	Per cent pentene disproportionation over	
	$WO_3 \cdot SiO_2$	$WO_3 \cdot SiO_2/MgO$
100	17.9	28.4
200	21.4	33.9

[a] Other conditions: 371 °C and 49–52 WHSV.

III. Homogeneous Disproportionation Catalysts

In 1967, Calderon, Chen, and Scott [4] reported a homogeneous catalyst system comprised of tungsten hexachloride, ethanol, and ethylaluminum dichloride would disproportionate internal olefins. These authors used the term "olefin metathesis" to describe the reaction. At room temperature 2-pentene was transformed in one to three minutes into a mixture containing, at equilibrium, 25, 50, and 25 mole per cent of 2-butene, 2-pentene, and 3-hexene, respectively. Double-bond isomerization was not detected and a quantitative reaction selectivity was obtained. Additional reports by Calderon and coworkers[45, 46]) disclosed that a wide range of Al:W ratios could be tolerated and a catalyst atomic ratio of 4:1:1 for Al:W:O was consistently very active.

To minimize possible cationic reactions due to aluminum compounds Wang and Menapace [47]) used *n*-butyl lithium as the reducing agent for tungsten hexachloride. With *cis*-2-pentene and mixture of *trans*- and *cis*-2-pentene, equilibrium conversions (50 per cent) were reached in four hours; however, under the same conditions, conversion of *trans*-2-pentene was only 40 per cent. Regardless of the extent of conversion of 2-pentene, the selectivity to 2-butene and 3-hexene was 100 per cent in experiments at room temperature with ratio of olefin/W = 50 and *n*-Bu Li/W = 2.

Combinations of nitrosyl molybdenum and tungsten compounds with organo-aluminum halides as active homogeneous disproportionation catalyst systems were disclosed by Zuech [5]) and reported in detail by Zuech and co-workers [48−52]). Treatment of green nitrosyl complexes, $L_2 Cl_2 (NO)_2 M$ [M = Mo or W; L = $Ph_3 P$, $C_5 H_5 N$, $Ph_3 PO$, etc.] with a variety of alkylaluminum halides in chlorobenzene yielded brown homogeneous solutions, which were very active at $0 - 50°$ C for disproportionating terminal as well as internal olefins. The versatility of the catalysts was demonstrated by reaction of alpha olefins, internal olefins, diolefins, and cleavage of both acylic and cyclic olefins with ethylene.

IV. Reactions and Reactants

Disproportionation reactions of acyclic mono-olefins can be classified into three types: (I) reactant is a single olefin, (II) reactants are double-bond isomers, and (III) reactants are different olefins. Generalized equations and examples are:

Type I   2 A $\rightleftarrows$ P + Q

                       2 1-Butene $\rightleftarrows$ Ethylene + 3-Hexene

Type II   A + A′ $\rightleftarrows$ P + Q

          1-Butene + 2-Butene $\rightleftarrows$ Propylene + 2-Pentene

Certain reactants yield two sets of products and in these cases the total primary product will consist of a mixture of four olefins. For example:

(a + b) 2-Methyl-2-butene + (a + b) 2-Methyl-1-butene $\rightleftarrows$

a (Isobutene + 3-Methyl-2-pentene) + b (Propylene + 2,3-Dimethyl-2-pentene)

Type III

$$A + B \quad \rightleftharpoons \quad P + Q \text{ (or to 2P)}$$

1-Butene + 2-Pentene $\rightleftharpoons$ Propylene + 3-Hexene

Ethylene + 2-Pentene $\rightleftharpoons$ Propylene + 1-Butene

Ethylene + 2-Butene $\rightleftharpoons$ 2 Propylene

As in the case with some Type II reactants, certain Type III reactants yield two sets of products. For example:

(a + b) Propylene + (a + b) 2-Methyl-2-pentene $\rightleftharpoons$

a (1-Butene + 2-Methyl-2-butene) + b (Isobutene + 2-Pentene)

Reverse of Type I and II reactions are Type III reactions. As stated in the introduction, abundant experimental data and mechanistic studies indicate that these catalytic disproportionation reactions proceed in accordance with a reaction scheme that can be pictured as follows:

$$
\begin{matrix}
R_2 & R_3 \\
| & | \\
R_1-C=C-R_4 \\
R_5-C=C-R_8 \\
| & | \\
R_6 & R_7
\end{matrix}
\rightleftharpoons
\begin{bmatrix}
R_2 & R_3 \\
| & | \\
R_1-C-C-R_4 \\
| & | \\
R_5-C-C-R_8 \\
| & | \\
R_6 & R_7
\end{bmatrix}
\rightleftharpoons
\begin{matrix}
R_2 & R_3 \\
| & | \\
R_1-C \quad C-R_4 \\
\| \quad \| \\
R_5-C \quad C-R_8 \\
| & | \\
R_6 & R_7
\end{matrix}
$$

where the R's are hydrocarbon groups or hydrogen and the brackets indicate a transition state. According to this scheme, the products of Type I reactions, as illustrated for disproportionation of 1-butene, will be symmetrical internal olefins of longer and shorter chain lengths.

$$
\begin{matrix}
C=C-C-C \\
C=C-C-C
\end{matrix}
\rightleftharpoons
\begin{bmatrix}
C-C-C-C \\
| \quad | \\
C-C-C-C
\end{bmatrix}
\rightleftharpoons
\begin{matrix}
C \quad C-C-C \\
\| \quad \| \\
C \quad C-C-C
\end{matrix}
$$

The transition state formed by opposite alignment of the reacting molecules disassociates into product molecules that are the same as the reacting olefins.

$$
\begin{matrix}
C=C-C-C \\
C-C-C=C
\end{matrix}
\rightleftharpoons
\begin{bmatrix}
C-C-C-C \\
| \quad | \\
C-C-C-C
\end{bmatrix}
\rightleftharpoons
\begin{matrix}
C \quad C-C-C \\
\| \quad \| \\
C-C-C \quad C
\end{matrix}
$$

In cases where the reactant is a symmetric olefin, e.g. 2-butene, the product will also be the same as the reactant.

$$C-C=C-C \\ C-C=C-C \quad \rightleftharpoons \quad \begin{bmatrix} C-C-C-C \\ | \quad | \\ C-C-C-C \end{bmatrix} \quad \rightleftharpoons \quad \begin{matrix} C-C \quad C-C \\ \| \quad \| \\ C-C \quad C-C \end{matrix}$$

However, mixtures of symmetrical and nonsymmetrical double-bond isomers, e.g., 2-butene and 1-butene, will be Type II reactions and the product will be longer and shorter chain olefins.

$$C-C=C-C \\ \quad C=C-C-C \quad \rightleftharpoons \quad \begin{bmatrix} C-C-C-C \\ | \quad | \\ C-C-C-C \end{bmatrix} \quad \rightleftharpoons \quad \begin{matrix} C-C \quad C-C \\ \| \quad \| \\ C \quad C-C-C \end{matrix}$$

Statistically, the rate of formation of new olefins for Type II reactions, except in some cases involving branched olefins, is twice the rate for Type I reactions; both alignments of the double-bond isomers to form the transition state will result in products different from the reactants. For Type III reactions three situations can exist; depending on the reactants, both, only one, or neither alignment of the olefins will form a four-center transition state that dissociates into new olefins. Cleavage of 2-butene with propylene will not form new olefines; however, cleavage of 2-butene with ethylene will form propylene.

$$C-C=C-C \\ \quad C=C-C \quad \rightleftharpoons \quad \begin{bmatrix} C-C-C-C \\ | \quad | \\ C-C-C \end{bmatrix} \quad \rightleftharpoons \quad \begin{matrix} C-C \quad C-C \\ \| \quad \| \\ C \quad C-C \end{matrix}$$

$$C-C=C=C \\ \quad C=C \quad \rightleftharpoons \quad \begin{bmatrix} C-C-C-C \\ | \quad | \\ C-C \end{bmatrix} \quad \rightleftharpoons \quad \begin{matrix} C-C \quad C-C \\ \| \quad \| \\ C \quad C \end{matrix}$$

Mixtures of acyclic and cyclic olefins apparently react via the four-center scheme to produce acyclic diolefins [39, 58]. These can be classified as Type IV reactions. In cases involving ethylene as the acyclic reactant, a, ω-diolefins are obtained.

Type IV

$$C + A \rightleftharpoons S$$

Cyclopentene + Ethylene $\rightleftharpoons$ 1,6-Heptadiene

The reactions of cyclic olefins by the four-center reaction scheme to form larger ring cyclic polyolefins and interlocked ring systems (catenanes) have also been proposed [59-67]. These can be classified as Type V reactions.

Type V

$$2\,C \rightleftharpoons T$$

2-Cyclopentene $\rightleftharpoons$ Cyclodecadiene

Presented in Table 6 are examples from the literature of Type I reactions. Again it should be noted that these data do not necessarily indicate optimum conditions for each catalyst or reactant. Also, in some cases double-bond shift occurred followed by Type II disproportionation reaction.

Table 6. *Examples of selective disproportionation of monoolefins:*
Type I reactions $2\,A \rightleftharpoons P + Q$

Olefin Reactant	Catalyst	Conv., %	Primary products Olefin	Mole %	Select. %	Ref.
Propylene	$Mo(CO)_6 \cdot Al_2O_3$	25	Ethylene 2-Butene	42 55	97	1)
	$CoO \cdot MoO_3 \cdot Al_2O_3$	36.1	Ethylene n-Butenes	49.0 48.5	97.5	1)
	$Re_2O_7 \cdot Al_2O_3$	31.7	Ethylene 2-Butene	55.5 44.5	100	11)
	$WO_3 \cdot SiO_2$	44.8	Ethylene n-Butenes	52.4 46.3	98.7	16)
1-Butene	$CoO \cdot MoO_3 \cdot Al_2O_3$ + Na	12.1	Ethylene Hexenes	42.3 38.4	80.7	37)
	$Re_2O_7 \cdot Al_2O_3$	37.9	Ethylene Hexenes	58.5 37.0	95.5	53)
Isobutene	$WO_3 \cdot SiO_2$ + Na	11	Ethylene Dimethylbutene } 88		88	38)
1-Pentene	$CoO \cdot MoO_3 \cdot Al_2O_3$	8.6	Ethylene n-Octenes	32.4 31.0	63.4	54)
	$(L)_2Cl_2(NO)_2Mo$ + $R_3Al_2Cl_3$	52	Ethylene 4-Octene	52 48	100	48)
2-Pentene	$CoO \cdot MoO_3 \cdot Al_2O_3$	35.3	n-Butenes n-Hexenes	47 36	83	3)
	$CoO \cdot MoO_3 \cdot Al_2O_3$	14.1	n-Butenes n-Hexenes	48.3 44.0	92.3	54)

Table 6 (continued)

Olefin Reactant	Catalyst	Conv., %	Primary products Olefin	Mole %	Select. %	Ref.
	$WCl_6 \cdot EtOH \cdot EtAlCl_2$	49.9	2-Butene 3-Hexene	49.8 50.2	100	4)
	$WCl_6 \cdot$ n-BuLi	50	2-Butene 3-Hexene	49 51	100	47)
	$(L)_2Cl_2(NO)_2Mo + RAlCl_2$	50	2-Butene 3-Hexene	48 52	100	55)
2-Hexene	$WCl_6 \cdot EtOH \cdot EtAlCl_2$	49	2-Butene 4-Octene	46 54	100	46)
2-Heptene	$WCl \cdot EtOH \cdot EtAlCl_2$	43	2-Butene 5-Decene	48 52	100	46)
3-Heptene	$MoO_3 Al_2O_3$ + K	71.8	3-Hexene 4-Octene	45.6 45.6	91.2	9)
1-Octene	$CoO \cdot MoO_3 \cdot Al_2O_3$ + Rb	70	Ethylene Tetradecene	41.6 42.1	83.7	14)
2-Octene	$MoO_3 \cdot Al_2O_3$+K	71.9	2-Butene 6-Dodecene	47.6 44.3	91.9	9)
	$WCl_6 \cdot EtOH \cdot EtAlCl_2$	48	2-Butene 6-Dodecene	47 53	100	46)

Literature examples of Type II reactions, disproportionation of double-bond isomers, are shown in Table 7. In addition to the Type II reaction, reaction of each nonsymmetrical olefin with itself can also occur resulting in broader distribution of products. A similar situation of Type I and II reactions proceeding simultaneously occurs for single olefin feed in cases where the catalyst system has significant double-bond isomerization activity.

Table 7. *Examples of olefin disproportionation reactions:*
Type II $A + A \rightleftarrows P + Q$

Reactants	1-Butene 2-Butene	51 49	1-Butene 2-Butene	20 80	n-Pentenes[a]	n-Heptenes[a]
Catalyst	$CoO \cdot MoO_3 \cdot Al_2O_3$		$CoO \cdot MoO_3 \cdot Al_2O_3$		$WO_3 \cdot SiO_2$[b]	$WO_3 \cdot SiO_2$[b]
Products, Mole %						
Ethylene	4		1			c)
Propylene	46		51		9	1
Butenes	–		–		41	11
Pentenes	43		44		–	18

R. L. Banks

Table 7 (continued)

Reactants	1-Butene 51 / 2-Butene 49	1-Butene 20 / 2-Butene 80	n-Pentenes[a]	n-Heptenes[a]
Hexenes } 7		4	32	24
Heptenes }			11	–
Octenes			6	21
Nonenes			1	14
Decenes			c)	7
C_{11}^+				4
Conversion, %	30	36	65	74
Ref.	44)	56)	15)	15)

a) Substantially equilibrium mixtures of n-olefins at 400 °C.
b) Catalyst pretreated with carbon monoxide.
c) Trace.

Examples of the disproportionation of dissimilar olefins, Type III reactions, are shown in Table 8. Alpha olefins were produced by disproportionating

Table 8. *Disproportionation of dissimilar olefins:*
Type III reactions A + B $\rightleftharpoons$ P + Q

Reactants	Propylene 58 / Isobutene 42	2-Butene 41 / Isobutene 59	Ethylene 50 / 2-Butene 50	Ethylene 87 / 2-Octene 13	Ethylene 70 / 4-MP-2[e] 29
Catalyst	$WO_3 \cdot SiO_2$	$WO_3 \cdot SiO_2$	$Re_2O_7 \cdot Al_2O_3$	$MoO_3 \cdot Al_2O_3$ b)	$CoO \cdot MoO_3 \cdot Al_2O_3$
Product, Mole %					
Ethylene	48	7	–	–	–
Propylene	–	42	97	2c)	37
Butenes	22a)	–	–	10	13
Isopentenes	26	41	–	–	37f)
n-Pentenes	1	1	3	3	2
Hexenes				Trace	6
Heptenes } 3	} 9		82d)		
Octenes				} 5	
C_9^+				3	
Conversion, %	36	57	53g)	29g)	63g)
Ref.	57)	57)	56)	9)	8)

a) n-Butenes.
b) Catalyst treated with potassium hydroxide.
c) Most of the propylene escaped from system.
d) The heptene was 98.4 per cent 1-heptene.
e) 4-Methyl-2-pentene.
f) The isopentene was 82 per cent 3-methyl-1-butene.
g) Conversion of the higher olefin reactant.

52

ethylene with higher olefins. With catalyst systems of sufficient double-bond isomerization activity, exhaustive ethylene cleavage of linear olefins to propylene and of olefins containing isolated methyl branches to isobutene and propylene have been demonstrated [44].

Presented in Table 9 are examples of Type IV reactions, cyclic monoolefins reacting with acyclic olefins. 1,6-Heptadiene, 1,7-octadiene, and 1,9-decadiene were produced when cyclopentene, cyclohexene, and cyclooctene were cleaved with ethylene. A tridecadiene was synthesized from cyclopentene and 1-octene.

Table 9. *Disproportionation of cyclic-acyclic olefins:* [58]
Type IV reactions $C + A \rightleftharpoons S$

Reactants	Product	Catalyst[a]	Conversion, %[b]	Selectivity, %
Cyclopentene Ethylene	1,6-Heptadiene	A	15–20	~100
Cyclopentene 1-Octene	Tridecadiene	A	5–10	~75
Cyclohexene Ethylene	1,7-Octadiene	B	–	36
Cyclooctene Ethylene	1,9-Decadiene	B	31	66

[a] Catalyst A: $Mo(CO)_6 \cdot Al_2O_3$
Catalyst B: $CoO \cdot MoO_3 \cdot Al_2O_3$

[b] Conversion of the cyclic olefin.

Examples of cyclic olefins forming larger ring polyolefins when contacted with disproportionation catalysts, Type V reactions, have been reported. Cyclohexadecadiene was produced when a dilute solution of cyclooctene was contacted with rhenium oxide-aluminum catalyst by workers at British Petroleum Company [59]. A cobalt molybdate catalyst was used by workers at Shell International [60] to convert cyclooctene to a series of oligomeric monocyclo-olefins. Reactions of cyclic olefins over the $WCl_6 \cdot EtAlCl_2 \cdot EtOH$ soluble disproportionation catalyst were reported by Goodyear workers [61–64] suggesting that ring-opening polymerization of cyclic olefins proceeded by the four-center reaction scheme. Wasserman and coworkers [65] identified cycloolefin rings in the product obtained by reacting cyclooctene in the presence of this soluble disproportionation catalyst system. Wolovsky [66] reported that a mixture of interlocked ring systems in the disproportionation product of cyclododecene could be identified by mass spectroscopic analysis. In an accompanying communication, Wasserman and coworkers [67] provide independent verification of the suggestion that cyclic olefins should yield catenated systems under disproportionation conditions.

Olefin disproportionation reactants are not limited to monoolefins; diolefins have been shown to react with another diolefin or with a monoolefin. Experiments were reported by Heckelsberg, Banks, and Bailey [68] in which 1,3-butadiene disproportionated to ethylene and cyclohexadiene, an isomer of hexatriene; 1,3-butadiene plus propylene to ethylene and 1,3-pentadiene; 1,3-pentadiene plus isobutene to ethylene and 4-methyl-1,3-pentadiene; and isoprene plus 2-butene to produce some 2-methyl-1,3-pentadiene and 3-methyl-1,3-pentadiene. The catalyst used in these experiments was tungsten oxide-silica that had been treated with sodium carbonate solution to reduce isomerization and polymerization side reactions. Ray and Crain [58] cleaved 1,5-cyclooctadiene with ethylene to yield 1,5,9-decadiene.

The disproportionation of acetylenes has also been reported: Pennella, Banks, and Bailey [69] found that 2-pentyne could be converted to 2-butyne and 3-hexyne when contacted with tungsten oxide-silica. The results were consistent with a four-center reaction scheme; the mole ratio of 3-hexyne to 2-butyne was always 1.0 to 1.2 (theoretical 1.0) and no significant amounts of other products with carbon numbers less than 10 were found.

V. Mechanism and Kinetics

Product Composition Studies

Banks and Bailey [1] reported olefin disproportionation products of a single olefin reactant contained approximately equal molar quantities of olefins having chains shorter and longer than the reactant and did not contain significant amounts of dimer. They concluded that two olefin molecules were reacting to form two olefin molecules. These authors noted at 204 °C propylene conversion was near the apparent equilibrium for the reaction

$$\text{2 Propylene} \rightleftharpoons \text{Ethylene} + \text{2-Butene}$$

and that the addition of ethylene or 2-butene to the propylene feed decreased propylene conversion. Efficiencies of converted propylene to ethylene and 2-butene were highest at low temperatures, and decreasing contact time (increasing space rate) decreased the fractions of 1-butene and of C_5^+ material in the product (Fig. 1). With higher olefin feeds approximately equal molar quantities of shorter and longer chain olefins were obtained, but the products were distributed nonselectively.

A detailed study of the disproportionation of 1-butene by Bradshaw, Howman, and Turner [8] showed that with reduced temperature or increased space

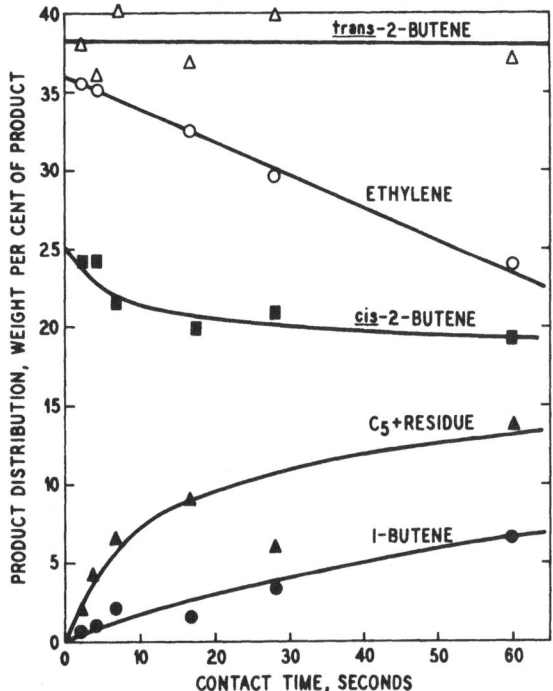

Fig. 1. Relation of product distribution to calculated contact time in propylene disproportionation. Data obtained in tests at 163 °C and 450 p.s.i.g. with CoO-MoO₃-Al₂O₃
catalyst and 60 propylene-40 propane feed (Ref. [1])

rate the amount of isomerization of 1-butene to 2-butene decreased and the
yield of ethylene and 3-hexene increased. The effect of double-bond isomerization on selectivity to ethylene and 3-hexene was further demonstrated with
a series of catalysts prepared by treatment of cobalt-molybdate with various
amounts of sodium bicarbonate to poison the isomerization sites. These workers suggested that the reaction occurs via a "quasi-cyclobutane" intermediate formed by the correct alignment of the carbon atoms at the double bonds
of two reacting olefins. Applied to 1-butene, the reaction was pictured as
follows:

$$
\begin{array}{ccc}
\mathrm{C{=}C{-}C{-}C} & \mathrm{C{\cdots}C{-}C{-}C} & \mathrm{C \quad C{-}C{-}C} \\
\rightleftharpoons & \rightleftharpoons & \| \; + \; \| \\
\mathrm{C{=}C{-}C{-}C} & \mathrm{C{\cdots}C{-}C{-}C} & \mathrm{C \quad C{-}C{-}C}
\end{array}
$$

Studies with ethylene plus 2-butene and with ethylene plus 4-methyl-2-
pentene provided additional support for this scheme and demonstrated that
disproportionation reactions are reversible.

Other investigators have made studies with heterogeneous catalysts under conditions that give high selectivities and reported product distributions that are consistent with the four-center mechanism. Crain [9] showed that over a potassium hydroxide-treated molybdena-alumina catalyst, 3-heptene dispro-portionates to 3-hexene and 4-octene, 1-octene to ethylene and 7-tetradecene, and 2-octene to 2-butene and 6-dodecene at selectivities greater than 80 per cent. He demonstrated that cleavage of 2-octene with ethylene gives high selectivity to 1-heptene and cleavage of 2,3-dimethylbutene-2 with ethylene gives isobutene. In this last reaction, Crain did not detect methylbutenes, pro-pylene or n-butenes and concluded that formation of the cyclobutadiene in-termediate, suggested by Mol and coworkers [70], was a remote possibility (this case requires the migration of two methyl groups and two hydrogen atoms to the catalyst surface as well as their return to the same carbon atom on which they were originally located). Adams and Brandenberger [54] conduc-ted studies with n-pentenes and 1-pentene/1-hexene and concluded that the mechanism of olefin disproportionation consisted of the reaction of two ole-fins with the transition metal oxide catalyst, followed by a concerted cyclo-butanation of the ligand-bound olefins and decomposition of the absorbed cy-clobutene to give product olefins with conservation of molecular orbital symmetry. Banks and Regier [57] demonstrated that isobutene plus propylene or 2-butene could be disproportionated at high selectivity to products con-sistent with the four-center reaction scheme.

A similar mechanism scheme was suggested by Calderon, Chen, and Scott [4] for olefin metathesis catalyzed by a homogeneous complex from the inter-action of WCl_6, C_2H_5OH, and $C_2H_5AlCl_2$. They proposed a transalkylidena-tion reaction as follows:

$$
\begin{array}{cc}
R_1 - CH \, \vdots \, CH - R_2 & R_1 - CH \quad CH - R_2 \\
\rightleftharpoons & \| \quad + \quad \| \\
R_1 - CH \, \vdots \, CH - R_2 & R_1 - CH \quad CH - R_2
\end{array}
$$

In subsequent publication, Calderon and coworkers [45] pictured the trans-sition state as

where W* denotes the transition metal atom plus the remaining ligands. The product distribution reported by these workers as well as those reported for other homogeneous disproportionation catalyst systems [47, 48] are consistent the four-center reaction scheme.

Carbon-14 Studies

Studies on the mechanism of disproportionation using ^{14}C-labeled propylene have been reported by several groups of investigators [70−73, 31]. Mol, Moulijn, and Boelhouwer reacted propylene labeled with ^{14}C in each of the three positions over rhenium oxide-alumina catalyst at moderate temperatures [70, 73]. The data were extrapolated to zero contact times to eliminate the influence of isomerization reactions. With [2-^{14}C] propylene the ethylene formed showed no radioactivity at all, in contrast with the butenes, which showed a specific radioactivity twice as high as that of the starting material. Experiments with [1-^{14}C] propylene and [3-^{14}C] propylene showed that the two methyl groups retain their identity throughout the disproportionation. These experiments support a four-center mechanism and, according to the authors, exclude the possibility of both a linear mechanism and a π-allylic intermediate in which the end carbon of propylene became indistinguishable.

Clark and Cook [71] disproportionated [1-^{14}C] propylene and [2-^{14}C] propylene over cobalt oxide-molybdate-alumina catalyst. At 60 °C their results were consistent with those reported Mol and coworkers, confirming the four-center mechanism. At temperatures above 60 °C, double-bond isomerization activity of the cobalt-molybdate catalyst became a factor and at 160 °C nearly one-half of [1-^{14}C] propylene had isomerized to [3-^{14}C] propylene prior to disproportionation. The authors note that at temperatures where isomerization does not occur, the possibility of a π-allyl intermediate appears to be excluded; however, at higher temperatures, the π-allyl mechanism cannot be so easily dismissed.

Woody, Lewis, and Wills [72] studied the disproportionation of [1-^{14}C] propylene over cobalt oxide-molybdate-alumina at 149 and 177 °C. Approximately equal amounts of radioactivity were found in the approximately equal molar quantities of ethylene and butene. These results are in agreement with those of Clark and Cook showing that double-bond isomerization was a factor in this temperature region. Woody and coworkers suggest that since the isomerization of the 2-butene product was negligible, an explanation of double-bond mobility as simple isomerization is probably an oversimplification.

Isagulyants and Rar [31] used radioactive propylene for investigating disproportionation on alumina. Their studies confirmed a mechanism of this reaction which includes the formation of an intermediate four-member surface ring.

Deuterated Olefins Studies

Calderon et al. [45] reacted 2-butene-d_8 with 2-butene and with 3-hexene in the presence of the homogeneous tungsten complex and obtained disproportionation product with mass consistent with transalkylidenation or the four-center scheme; with 2-butene, the only new olefin had mass corresponding to $C_4 H_4 D_4$, and with 3-hexene, corresponding to $C_5 H_6 D_4$. These results were used to eliminate a transalkylation scheme, involving the interchange of alkyl groups via scission alpha to the double-bond, which had been considered by these authors.

Mol, Visser, and Boelhouwer [74] passed 2-deuteropropylene over rhenium oxide-alumina catalyst at 85 °C to obtain information concerning the mobility of ring hydrogen atoms during the disproportionation reaction. Mass spectrometry showed that the ethylene product was nearly free of deuterium and that the butene contained two deuterium atoms. It was concluded that no hydrogen exchange takes place during disproportionation. The authors noted that their results are in correspondence with unpublished results of Olsthoom, who subjected a mixture of $C_2 H_4$ and $C_2 D_4$ to disproportionation conditions and obtained besides the starting material only $C_2 H_2 D_2$, indicating that a cyclobutane structure plays a part in the reaction mechanism.

Experiments with Cyclobutanes

Mol, Visser, and Boelhouwer [74] subjected 1,2-dimethyl-butane (ring structure of the suggested transition state for disproportionation of propylene) to rhenium oxide-alumina catalyst under conditions which propylene gives high disproportionation conversions. This compound was stable; only at high temperatures (730 °C), where thermal cracking occurred, were olefins found.

Pettit, Sugahara, Wristers, and Merk [75] found that the gas phase reaction of tetramethylcyclobutane with molybdenum on alumina readily yield 2-butene and subsequent products derived therefrom. No reaction occurred when cis-tetramethylcyclobutane was treated with tungsten containing homogeneous disproportionation catalyst. These authors believe that the disproportionation of olefins does involve an allowed metal participating cycloaddition reaction of two ethylenic units and the failure of the cyclobutane to cleave in the presence of the homogeneous catalyst resulted from the inability to position the metal atom and the saturated hydrocarbon in close proximity.

Stereochemistry Studies

Hughes [76] studied the stereochemistry of the homogeneous disproportionation of 2-pentene using soluble molybdenum catalyst system. His results reveal that the molybdenum catalyst exhibits a high degree of stereoselectivity; cis-2-pentene disproportionated preferentially to cis-2-butene and cis-3-hexene, and trans-2-pentene reacted to yield preferentially trans-2-butene and *trans*-3-hexene. He suggested a reaction intermediate involving a *cis*-diolefin-molybdenum complex.

Catenanes Synthesis

Independent interpretation of mass spectroscopic data by Wolovsky [66] and Ben-Efraim, Batich, and Wasserman [67] indicate that cyclic olefins subjected to disproportionation conditions form interlocking rings (catenanes) as well as cyclopolyolefins. The proposed scheme involves four-center transition state and "Mobius-Strip" approach

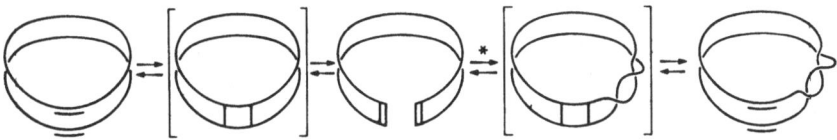

* 360 ° twist

A 180 ° twist would correspond to a Mobius strip which could form a single large ring; a 540 ° twist would yield a trifold knot and 720 ° twist a double treaded catenane.

Other Reaction Via Four-Center Scheme

Banks and Bailey [77] noted that the occurence of disproportionation, polymerization, and isomerization over similar catalysts, or simultaneously over the same catalyst, suggests a similarity of mechanism. They noted that this is not to say that one can predict a given catalyst will promote one or more of these reactions or that a given catalyst known to promote one of these reactions also will promote another. Banks and Bailey [77] proposed that the ability of the catalyst to shift hydrogen atoms is a key factor in determining the reaction course. When they contacted ethylene with a series of catalysts prepared by supporting Group VI hexacarbonyls on alumina, they obtained different products with the different hexacarbonyls (Table 10).

Table 10. *Products obtained from ethylene over group VI metal hexacarbonyls*

Catalyst: Alumina Impregnated with	Products
$W(CO)_6$	21 % Propylene, 71 % 1-butene, 8 % 2-butene
$Mo(CO)_6$	8 % Cyclopropane, 12 % methylcyclopropane, 28 % propylene, 26 % 1-butene, 26 % 2-butene
$Cr(CO)_6$	3 % Butenes, 97 % solid polyethylene

They proposed that 1-butene is obtained from ethylene by intermediate (I) in which breaking one bond and hydrogen shift occurs. 2-Butene is produced by double-bond isomerization, and propylene is formed from ethylene and 2-butene by breaking ring bonds C_1-C_2 and C_3-C_4 of structure (II).

$$\begin{array}{ccc}
C_1 \text{---} C_2 & C_1 \text{---} C_2 & C_1 \text{---} C_2 \\
\mid \quad \mid & \mid \quad \mid & \mid \quad \mid \\
C_4 \text{---} C_3 & C - C_4 \text{---} C_3 - C & C_4 \text{---} C_3 - C \\
\text{(I)} & \text{(II)} & \text{(III)}
\end{array}$$

Some catalyst sites are apparently specific enough to close the cyclopropane ring by breaking adjacent bonds in structure (II) and (III) producing methylcyclopropane and cyclopropane. Long chain polyethylene is produced by the same mechanism postulated for 1-butene if the product remains on the catalyst and reacts repeatedly with ethylene by the following sequence.

$$\begin{array}{ccc}
O \quad O & C_5 \text{---} C_6 & C_5 - C_6 - R' \\
\mid \quad \mid & \mid \quad \mid & \mid \quad \mid \\
\mid \quad \mid \; \rightarrow & \mid \quad \mid \; \rightarrow & \mid \quad \mid \\
R - C_4 \text{---} C_3 & R - C_4 \text{---} C_3 & O \quad O \\
\text{(IV)} & \text{(V)} & \text{(VI)}
\end{array}$$

Banks and Bailey concluded that disproportionation occurs when two molecules are adsorbed with the breakage of two opposite bonds without hydrogen shift. Polymerization occurs when two molecules are adsorbed with the formation of a four-center complex and then desorb with the breakage of one bond and hydrogen shift. Skeletal isomerization occurs when one molecule is adsorbed with the formation of a four-center complex and then desorbed with the breakage of one bond and hydrogen shift.

Infrared Spectral Studies

Infrared spectral studies on molybdenum hexacarbonyl-alumina were reported by Davie, Whan, and Kemball [78]. Without any activation procedure they obtained a sharp carbonyl frequency corresponding to unchanged hexacarbonyl on the support. This material was not active for disproportionation. After treatment for one hour under vacuum at 373 °K the catalyst had lost the sharp carbonyl band but showed two wider and broader bands and was active for disproportionating propylene. The authors stated that the active catalyst clearly had a lower symmetry than the hexacarbonyl and must have lost one or more of the carbonyl groups. After exposure of the activated catalyst to air, the catalyst was inactive and showed no absorption in the carbonyl region.

Homogeneous and Heterogeneous Tungsten Catalysts

Herisson and coworkers [79] studied the reactions of acyclic olefins catalyzed by homogeneous and heterogeneous tungsten catalysts. They concluded that the presence of a Lewis-type acid appears to be indispensable, the degree of oxidation of the tungsten plays only a secondary role, and the homogeneous or heterogeneous nature does not appear to influence the course of the reaction.

Molecular Orbital Symmetry Considerations

In absence of a catalyst, simple olefins are essentially fixed in their bonding configurations; reaction paths to interconversions through molecular collisions, fusions, and disassociations are apparently closed because of orbital symmetry restrictions, as proposed by Hoffman and Woodward [80]. Mango [81] has postulated that in the presence of certain transition metal catalysts, these orbital symmetry restraints are lifted, allowing bonds to flow freely and molecular systems to interchange. Thus, the conservation of molecular orbital symmetry is a key function of the catalyst.

Kinetic Studies

Propylene disproportionation kinetics using tungsten oxide-silica catalyst were studied by Begley and Wilson [82] in an attempt to develop a kinetic model for design and optimization studies. The kinetics were examined in terms of Langmuir-Hinshelwood and Rideal models and the Rideal model was found to fit the data adequately when equivalent surface conditions existed, i.e., at pressures of 300–900 psig. They reported that at lower pressures, 50 and 115 psig, the rate constants were higher and suggested that this was because of a change in the linear gas velocity in the catalyst bed.

Lewis and Wills [83] obtained initial differential rate data for the disproportionation of propylene over a $CoO-MoO_3-Al_2O_3$ catalyst. Temperatures of 394-478 °K and pressures of 1 to 9 atmospheres were used. The authors reported the experimental data were well correlated when it was assumed that a dual site surface reaction was the controlling step in the mechanism.

Moffat and Clark [84] found that a Langmuir-Hinshelwood model applied to a heterogeneous surface can be used to describe both the general kinetics and the rate-temperature maxima reported by Banks and Bailey (Fig. 2) for olefin disproportionation on cobalt molybdate-alumina catalyst. They conclude that the rate-temperature maximum was caused by the reversible deactivation of sites superimposed on the irreversible poisoning of sites.

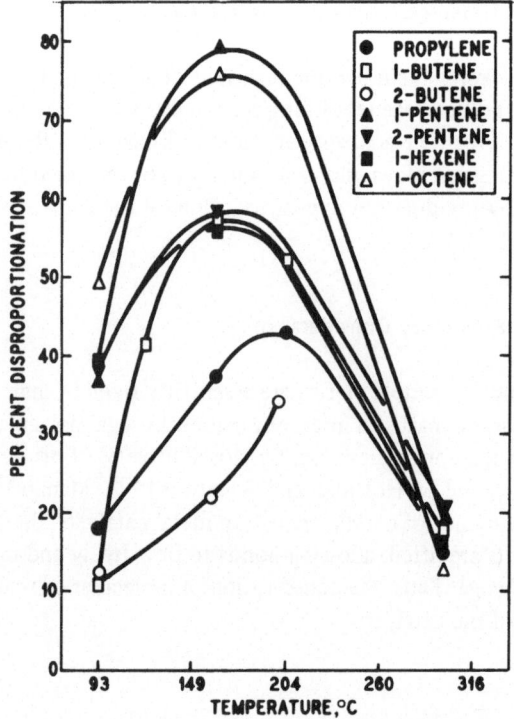

Fig. 2. Effect of temperature on disproportionation conversion (Cobalt molybdate – alumina catalyst) (Ref. [1])

Kinetics studies of the homogeneous disproportionation of 2-pentene to 2-butene and 3-hexene using a $L_2Mo(NO)_2Cl_2$-R_xAlCl_{3-x} catalyst was made by Hughes [55]. He showed the rate of disappearance of 2-pentene to be first order in catalyst and variable order in olefin. At low olefin:catalyst ratios the order is greater than one, at higher olefin:catalyst ratios the order is approximately one. These results are interpreted in terms of mechanism involving stepwise, rapid reversible olefin complexation followed by a rate-determining disproportionation step. Hughes suggests that the metal functions as a reaction template and in this manner overcomes the high entropy requirements of the reaction.

Activation Energy

The reported apparent activation energies for disproportionating propylene are lower for the cobalt molybdate-alumina than for tungsten oxide-silica. With cobalt molybdate-alumina an activation energy of 7.7 kcal/mole was obtained by Clark and Cook[71] and of 8.2 kcal/mole by Moffat and Clark[84]. Begley and Wilson[82] determined the activation energy for the reaction on an aged tungsten oxide-silica catalyst to be 21.6 kcal/mole and on a new catalyst, 18.6 kcal/mole. Activation energies of homogeneous disproportionation utilizing molybdenum nitrosyl catalysts were reported by Hughes[55] to be 6.6 and 7.0 kcal/mole for 2-pentene and 4-nonene, respectively.

Calculated Equilibrium Conversions

The calculated thermodynamic equilibrium conversions and product compositions for propylene disproportionation at 200 to 400 °C were reported by Heckelsberg, Banks, and Bailey[16]. Atlar, Pis'man, and Bakhshi-Zade[85] made similar calculations for the 50 – 300 °C range. They noted that the equilibrium constants were independent of pressure. Banks and Regier[57] showed thermodynamic equilibrium conversions as a function of temperature for the various reactions involved in the synthesis of isoamylene via disproportionation (Fig. 3). A comparison of calculated equilibrium composition for

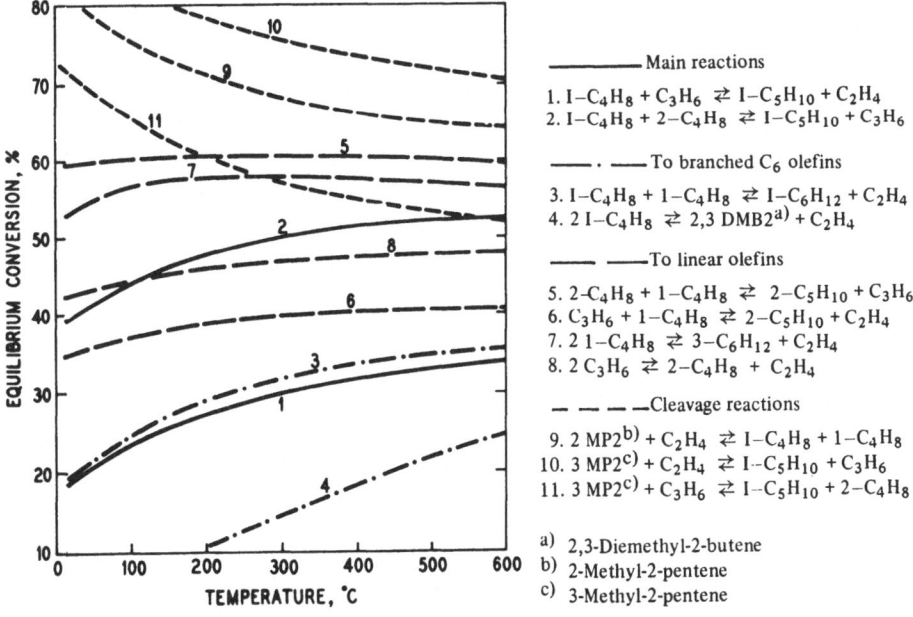

——————— Main reactions

1. $1-C_4H_8 + C_3H_6 \rightleftarrows 1-C_5H_{10} + C_2H_4$
2. $1-C_4H_8 + 2-C_4H_8 \rightleftarrows 1-C_5H_{10} + C_3H_6$

— · —— To branched C_6 olefins

3. $1-C_4H_8 + 1-C_4H_8 \rightleftarrows 1-C_6H_{12} + C_2H_4$
4. $2\ 1-C_4H_8 \rightleftarrows 2,3\ DMB2^{a)} + C_2H_4$

——— ——To linear olefins

5. $2-C_4H_8 + 1-C_4H_8 \rightleftarrows 2-C_5H_{10} + C_3H_6$
6. $C_3H_6 + 1-C_4H_8 \rightleftarrows 2-C_5H_{10} + C_2H_4$
7. $2\ 1-C_4H_8 \rightleftarrows 3-C_6H_{12} + C_2H_4$
8. $2\ C_3H_6 \rightleftarrows 2-C_4H_8 + C_2H_4$

— — — —Cleavage reactions

9. $2\ MP2^{b)} + C_2H_4 \rightleftarrows 1-C_4H_8 + 1-C_4H_8$
10. $3\ MP2^{c)} + C_2H_4 \rightleftarrows 1-C_5H_{10} + C_3H_6$
11. $3\ MP2^{c)} + C_3H_6 \rightleftarrows 1-C_5H_{10} + 2-C_4H_8$

a) 2,3-Dimethyl-2-butene
b) 2-Methyl-2-pentene
c) 3-Methyl-2-pentene

Fig. 3. Disproportionation reactions in isoamylene synthesis (Ref. [57])

63

disproportionating propylene in 60 propylene — 40 propane mixture (which assumes double-bond isomerization, but no secondary disproportionation reactions occur) with typical product distribution reported by Heckelsberg and coworker is shown in Table 11.

Table 11. *Comparison of experimental product distribution with calculated equilibrium data* [a]

	Calculated	equilbrium	data[a]	Experimental data
Temperature, °C	204	315	426	426
Product composition, wt. %				
Ethylene	9.3	9.7	10.0	9.5
Propane	40.0	40.0	40.0	40.0
Propylene	32.1	31.0	30.1	33.1
1-Butene	2.6	4.0	5.2	5.4
trans-2-Butene	10.4	9.4	8.5	6.7
cis-2-Butene	5.6	5.9	6.2	4.7
Pentenes and Heavier				0.6
Conversion, %	46.5	48.3	49.8	44.8
Efficiency, %				97.8

[a] Assumes double-bond isomerization, but no secondary disproportionation reactions. Ref. [16]

Mass Transfer Effects

Moffat, Johnson, and Clark [86] found the propylene disproportionation reaction on tungsten oxide-silica catalyst to be limited by interphase diffusional effects in spite of calculations which predict that no diffusional limitation should occur. They postulate that widely separated and very active sites could have their inherent activity limited by localized film diffusional effects which are functions of Reynolds and Schmidt numbers. Activity of cobalt molybdate-alumina was not limited by interphase or prediffusional effects.

VI. Process Application

The first commercial application of olefin disproportionation was in 1966 [87]; Shawinigan Chemicals Ltd. at the Varennes complex near Montreal, Quebec brought onstream the Phillips Triolefin Process [88] for converting propylene into polymerization-grade ethylene and high-purity butenes. Pilot plant development, reported by Johnson [89], showed that during a 20-hour test propylene conversion remained nearly constant at 43 per cent and efficiency of converted propylene to ethylene and n-butenes increased from 93 to 99 per cent.

A decrease in 1-butene yield indicated that double-bond isomerization activity of the catalyst decreased more rapidly than disproportionation activity. Patents to Wilson and Larson [90–92] relate to the start-up and operation of olefin disproportionation processes.

Incorporation of the Triolefin Process with naphtha cracking, described by Dixon, Hutto, Wilson, and Banks [93], permits ethylene yields of over 45 weight per cent of the naphtha charged as compared to about 36 weight per cent obtained in high-severity cracking units. One process disclosed by Hutto and coworkers [94] showed how naphtha cracking, disproportionation, and olefin dehydrogenation can be used to produce ethylene and butadiene. Gilliland, MacQueen, and Dixon [95] combined olefin disproportionation, naphtha cracking, olefin dehydrogenation, hydrotreating and aromatic extraction to produce ethylene, butadiene, benzene, and a high aromatic content carbon feed stock. Dixon [96] combines disproportionation steps with hydrogenation and naphtha cracking unit to convert propylene to ethylene.

High quality gasoline components can be obtained by combining disproportionation and alkylation processes. Logan and Banks [97] describe a combined process for disproportionating propylene to ethylene and 2-butene and for alkylating these products to diisopropyl and 2-butene alkylate. The combined alkylate has a leaded blending research octane number of about 110 as compared to 103 for direct propylene alkylation. Process features are disclosed in a patent issued to Dixon [98]. Banks, Hutson, and Logan [99] have shown how pentenes in cat cracker gasoline can be removed and utilized. In one scheme, the pentenes are disproportionated to higher olefins, which are returned to gasoline, and to lower olefins, which are alkylated; in a second scheme the pentenes are cleaved with ethylene to yield high-purity propylene and/or alkylation feed stock. The authors point out that reduction of pentene content from about 10 per cent to 0.5 per cent constitutes a substantial reduction in the gasoline smog forming potential. A process in which the undivided effluent from a propylene disproportionating zone is alkylated with isobutenes is described by Philipps [100].

Laboratory development of Triolefin Process technology for synthesizing isoamylene, an intermediate in polyisoprene production, was reported by Banks and Regier [57]. Isoamylene purity of 92 per cent and isoamylene yield of 1.0 pounds per pound of isobutene converted were obtained with feeds containing isobutene, propylene, and n-butenes. Isobutene converted to C_6+ byproduct was recovered by cleaving the C_6+ material with ethylene or propylene to yield butenes and pentenes. Process for producing isoprene from butene streams is the subject of a patent issued to McGrath and Williams [101].

The process includes steps for isomerizing 1-butene into 2-butene at temperatures below 0 °C over alumina treated with alkaline metal and for disproportionating the 2-butene with isobutene over rhenium heptaoxide on alkaline-treated alumina. With cobalt-molybdate catalyst, Sampson and Jackson [102] found oligomerization of the isobutene predominated and disproportionation was the minor reaction.

Processes utilizing disproportionation to convert propylene and butene into long chain linear olefins (e.g., detergent range olefins) are the subject of several patents. Sherk [103] uses more than one stage of olefin disproportionation to convert a feed of acyclic olefins into higher and lower molecular weight olefins. Kenton, Crain, and Kleinschmidt [7] combine a series of olefin reaction steps with removal of branched dimer by-product from the effluent of each step to produce a mixture of linear olefins having 11 to 15 carbon atoms. Marsheck [104] shows how to interrelate disproportionation and fractionation steps to convert one or more alkenes into one or more other olefins. Heckelsberg [105] combines selectively disproportionation and double-bond isomerization steps to convert propylene to 5-decene. Davison [106] combines disproportionation and ethylene polymerization to convert propylene to alpha olefins having 12 to 20 atoms per molecule. Albright [107] uses dimerization and disproportionation to convert an olefin to an heavier olefin. Stapp and Crain [108] synthesize 3-methyl-1-butene from propylene with a combination of propylene dimerization and ethylene cleavage.

Bourne and Metcalfe [109] convert olefins containing the group

$$
\begin{array}{c}
CH_3 \\
| \\
= CH \cdot CH \cdot CH_3
\end{array}
$$

to paraxylene by combining disproportionation and dehydrocyclization. In one example 4-methylpentene-2 was disproportionated over rhenium oxide-alumina catalyst to yield 2,5-dimethylhexene and 2-butene and the dimethyl-hexene was converted to paraxylene over a chromia on alumina dehydrocyclization catalyst. Dixon [110] integrates disproportionation and dehydrogenation.

Heckelsberg and Banks [111] combine a dehydrogenation catalyst (e.g., $Cr_2O_3 \cdot Al_2O_3$) and a disproportionation catalyst in a single reactor to convert paraffinic hydrocarbons to a plurality of olefinic hydrocarbons. Banks [112] discloses a process for converting propane to diisopropyl: Propane is cracked to produce an effluent stream comprising hydrogen, ethylene, and propylene; propylene is disproportionated to produce butenes and additional ethylene; butenes are hydroisomerized to isobutane utilizing hydrogen from the cracking step; and the isobutane and ethylene are alkylated to diisopropyl. Another patent to Banks [113] gives examples of converting ethylene to propylene over tungsten oxide-silica catalyst; apparently the ethylene dimerizes and the butene product reacts with ethylene to yield propylene [77].

VII. References

[1] Banks, R. L., Bailey, G. C.: Ind. Eng. Chem., Prod. Res. Develop. *3*, 170 (1964).
[2] – U. S. Patent 3,463,827; Belgian Patent 633,418.
[3] – U. S. Patent 3,261,879; Belgian Patent 620,480.
[4] Calderon, N., Chen, H. Y., Scott, K. W.: Tetrahedron Letters *34*, 3327 (1967).
[5] Zuech, E. A.: Chem. Commun. 1182 (1968).
[6] Schneider, V., Frolich, P. K.: Ind. Eng. Chem. *23*, 1405 (1931).
[7] Kenton, J. R., Crain, D. L., Kleinschmidt, R. F.: U. S. Patent 3,491,163.
[8] Bradshaw, C. P. C., Howman, E. J., Turner, L.: J. Catalysis 7, 269 (1967).
[9] Crain, D. L.: J. Catalysis *13*, 110 (1969).
[10] Bailey, G. C.: Catalysis Rev. *3* (1), 37 (1969).
[11] Turner, L., Howman, E. J., Williams, K. V.: British Patent 1,106,016.
[12] Heckelsberg, L. F., Banks, R. L., Bailey, G. C.: Ind. Eng. Chem., Prod. Res. Develop. *8*, 259 (1969)
[13] Turner, L., Bradshaw, C. P. C.: British Patent 1,103,976.
[14] Van Helden, R., Engel, W. F., Alkema, H. J., Krijger, P.: British Patent 1, 164, 687.
[15] Heckelsberg, L. F.: U. S. Patent 3,365,513.
[16] – Banks, R. L., Bailey, G. C.: Ind. Eng. Chem., Prod. Res. Develop., *7*, 29 (1968).
[17] – U. S. Patent 3,418,390.
[18] – U. S. Patent 3,340,322.
[19] Atlas, V. V., Pis'man, I. I., Bakshi-Zade, A. M.: Khimicheskaya Promyshlennost *45* (10), 734 (1969).
[20] Heckelsberg, L. F.: U. S. Patent 3,444, 262.
[21] Williams, K. V., Turner, L.: British Patent 1,116,243; U. S. Patent 3,485,889.
[22] Howman, E. J., Turner, L., Williams, K. V.: British Patent 1,106,015.
[23] – Bradshaw, C. P. C., Turner, L.: British Patent 1,105,564.
[24] Howman, E. J., Turner, L.: British Patent 1,089,956.
[25] – – Williams, K. V., Bradshaw, C. P. C.: British Patent 1,093,784.
[26] Bradshaw, C. P. C., Turner, L.: British Patent 1,121,806.
[27] Turner, L., Williams, K. V.: British Patent 1,096,250.
[28] – – British Patent 1,096, 200.
[29] Banks, R. L.: U. S. Patent 3,442,969.
[30] Heckelsberg, L. F.: U. S. Patent 3,395,196.
[31] Isagulyants, G. V., Rar, L. F.: Izv. Akad. Nauk SSSR, Ser. Khim. *6*, 1362 (1969).
[32] Banks, R. L.: U. S. Patent, 3,546,313.
[33] Lapidus, A. L., Isakov, Y. I., Avetisyan, R. V., Sendel, A. K., Minachev, K. M., Eidus, Y. T.: Izv. Akad. Nauk SSSR, Ser. Khim. *1*, 57 (1970).
[34] Howman, E. J., Turner, L.: British Patent 1,105,563.
[35] Reusser, R. E.: Belgian Patent 713,183.
[36] Lester, G. R.: U. S. Patent 3,471,586.
[37] Turner, L., Howman, E. J., Bradshaw, C. P. C.: British Patent 1,056,980.
[38] Heckelsberg, L. F.: Belgian Patent 713,184.
[39] Mango, F. D.: U. S. Patent 3,424,811.
[40] Pennella, F.: Belgian Patent 713,187; Italian Patent 829,071.
[41] Alkema, H. J., Van Helden, R.: British Patent 1,174,968; Belgian Patent 722,432.
[42] Heckelsberg, L. F.: Belgian Patent 713,185.
[43] Alkema, H. J., Van Helden, R.: British Patent 1,117,968.
[44] Banks, R. L., Kenton, J. R.: Belgian Patent 713,190.

R. L. Banks

45) Calderon, N., Ofstead, E. A., Ward, J. P., Judy, W. A., Scott, K. W.: J. Am. Chem. Soc. *90*, 4133 (1968).
46) – Chen, H. Y.: South African Patent 671,913.
47) Wang, J., Menapace, H. R.: J. Org. Chem. *33*, 3794 (1968).
48) Zuech, E. A., Hughes, W. B., Kubicek, D. H., Kittleman, K. T.: J. Am. Chem. Soc. *92*, 528 (1970).
49) – Belgian Patent 714, 621.
50) Kittleman, E. T., Zuech, E.A.: Belgian Patent 714, 622.
51) Zuech, E. A.: Belgian Patent 714, 623.
52) Hughes, W. B., Zuech, E. A.: Belgian Patent 714,624.
53) Turner, L., Howman, E. J., Bradshaw, C. P. C.: British Patent 1,054,864.
54) Adams, C. T., Brandenberger, S. G.: J. Catalysis *13*, 360 (1969).
55) Hughes, W. B.: J. Am. Chem. Soc. *92*, 532 (1970).
56) Turner, L., Bradshaw, C. P. C., Howman, E. J.: British Patent 1,064,829.
57) Banks, R. L., Regier, R. B.: Preprints Division Petroleum Chemistry, ACS, Vol. 15, No. 2, E 72 (1970).
58) Ray, G. C., Crain, D. L.: British Patent 1,163,657; Belgian Patent 694,420.
59) Turner, L., Bradshaw, C. P. C.: British Patent 1,105,565.
60) Shell International Research Maatschappij, N. V.: British Patent 1,118,517.
61) Calderon, N.: U. S. Patent 3,439,057.
62) – U. S. Patent 3,439,056.
63) Scott, K. W., Calderon, N., Ofstead, E. A., Judy, W. A., Ward, J. P.: 155th National Meeting American Chemical Society, April 1968, Abstract L 54.
64) Calderon, N., Ofstead, E. A., Judy, W. A.: J. Polymer Sci. A 1, *5*, 2209 (1967).
65) Wasserman, E., Ben-Efraim, D. A., Wolovsky, R.: J. Am. Chem. Soc. *90*, 3286 (1968).
66) Wolovsky, R.: J. Am. Chem. Soc. *92*, 2132 (1970).
67) Ben-Efraim, D. A., Batich, C., Wasserman, E.: J. Am. Chem. Soc. *92*, 2133 (1970).
68) Heckelsberg, L. F., Banks, R. L., Bailey, G. C.: J. Catalysis *13*, 99 (1969).
69) Pennella, F., Banks, R. L., Bailey, G. C.: Chem. Commun. 1548 (1968).
70) Mol, J. C., Moulijn, J. A., Boelhouwer, C.: J. Catalysis *11*, 87 (1968).
71) Clark, A., Cook, C.: J. Catalysis *15*, 420 (1969).
72) Woody, F. L., Lewis, M. J., Wills, G. B.: J. Catalysis *14*, 389 (1969).
73) Mol, J. C., Moulijn, J. A., Boelhouwer, C.: Chem. Commun. 633 (1968).
74) – Visser, F. R., Boelhouwer, C.: J. Catalysis *17*, 114 (1970).
75) Pettit, R., Sugahara, H., Wristers, J., Merk, W.: Discussions Faraday Soc. *47*, 71 (1969).
76) Hughes, W. B.: Chem. Commun. 431 (1969).
77) Banks, R. L., Bailey, G. C.: J. Catalysis *14*, 276 (1969).
78) Davie, E. S., Whan, D. A., Kemball, C.: Chem. Commun. 1430 (1969).
79) Herisson, J. L., Chauvin, Y., Phung, N. H., Lefebvre, G.: Compt. Rend. Acad. Sci. Paris, Ser. C *269* (13), 661 (1969).
80) Hoffmann, R., Woodward, R. B.: J. Am. Chem. Soc. *87*, 2046 (1965).
81) Mango, F. D.: Advances in Catalysis, Vol. 20, pp. 307-311. New York: Academic Press 1969.
82) Begley, J. W., Wilson, R. T.: J. Catalysis *9*, 375 (1967).
83) Lewis, M. J., Wills, G. B.: J. Catalysis *15*, 140 (1969).
84) Moffat, A. J., Clark, A.: J. Catalysis *17*, 264 (1970).
85) Atlas, V. V., Pis'man, I. I., Bakhski-Zade, A. M.: Neftepererabotka i Neftekhimiya *5*, 30 (1969).

86) Moffat, A. J., Johnson, M. M., Clark, A.: J. Catalysis *18*, 345 (1970).

87) Chemical Week, July 16, 1966, p. 77 and July 23, 1966, p. 70.

88) Phillips Petroleum Company: Hydrocarbon Processes *46* (11), 232 (1967).

89) Johnson, P. H.: VII. World Petroleum Congress Proc., Mexico City 1967, Vol. 5, p. 247; – Hydrocarbon Processes *4*, 149 (1967).

90) Wilson, R. T., Larson, L. G.: U. S. Patent 3,346,661.

91) – – U. S. Patent 3,365,513.

92) Larson, L. G., Wilson, R. T.: Belgian Patent 713, 182.

93) Dixon, R. E., Hutto, J. F., Wilson, R. T., Banks, R. L.: Oil Gas J. *65* (4), 98 (1967).

94) Hutto, J. F., Dixon, R. E., Davison, J. W., Ayres, C. A., Renberg, G. A., Mac Queen, D. K.: U. S. Patent 3,345,285.

95) Gilliland, R. E., MacQueen, D. K., Dixon, R. E.: U. S. Patent 3,281,351.

96) Dixon, R. E.: U. S. Patent 3,485,890.

97) Logan, R. S., Banks, R. L.: Oil Gas J. *66* (21), 131 (1968); Proc. API Ref. *48*, 984 (1968).

98) Dixon, R. E.: Canadian Patent 755,135.

99) Banks, R. L., Hutson, T., Logan, R. S.: Proc. API Preprint 02–70.

100) Phillips, J. E.: U. S. Patent 3,236,912.

101) McGrath, B.P., Williams, K. V.: British Patent 1,170,498.

102) Samspon, R. J., Jakson, D.: British Patent 1,171,970.

103) Sherk, F. T.: U. S. Patent 3,296,330.

104) Marsheck, R. M.: Belgian Patent 678,016.

105) Heckelsberg, L. F.: U. S. Patent 3,485,891.

106) Davison, J. W.: U. S. Patent 3,442,970.

107) Albright, M. A.: U. S. Patent 3,330.882.

108) Stapp, P. R., Crain, D. L.: U. S. Patent 3,457,320.

109) Bourne, K. H., Metcalfe, C. J. L.: U. S. Patent 1,145,503.

110) Dixon, R. E.: Belgian Patent 633,483.

111) Heckelsberg, L. F., Banks, R. L.: U. S. Patent 3,445,541.

112) Banks, R. L.: U. S. Patent 3,321,547.

113) – U. S. Patent 3,431, 316.

Received October 9, 1970

Homogene katalytische Hydrierung

mit Komplexverbindungen der 8. Gruppe

Prof. Dr. Walter Strohmeier

Institut für Physikalische Chemie der Universität Würzburg

Inhalt

I. Einleitung

Noch vor einem Jahrzehnt waren nur wenige homogene Hydrierungskatalysatoren bekannt. Die Entwicklung auf dem Gebiet der Komplexverbindungen zeigte jedoch, daß vor allem unter den Komplexverbindungen der Metalle der 8. Gruppe viele Verbindungen sind, welche bereits im Temperaturbereich von +20 bis 80 °C und Normaldruck von H_2 sehr gute Hydrierungskatalysatoren sind (7,63). Hier wird ein Überblick über den derzeitigen Stand der Kenntnisse gegeben. Weiterhin werden die Ergebnisse quantitativer Untersuchungen dazu benützt, gewisse Regeln und Arbeitshypothesen herauszuarbeiten, mit deren Hilfe es möglich wird, gezielt aktivere und selektivere Hydrierungskatalysatoren zu finden.

II. Hydrierungskatalysatoren der 8. Gruppe

Die typischen Hydrierungskatalysatoren der 8. Gruppe des Periodensystems, welche hier diskutiert werden sollen, sind dadurch gekennzeichnet, daß das Zentralatom M der Komplexverbindung ein Metall der 8. Gruppe ist, dessen Koordinationsstellen durch ungeladene Liganden L wie z.B. Phosphine $(PR_1R_2R_3)$, Arsine, SR_2 etc. und/oder geladene Liganden X (z.B. Halogene, CN etc.) besetzt sind. Als weitere Liganden Y werden der Hydridwasserstoff H sowie die NO- und die CO-Gruppe in die Betrachtung mit einbezogen. Die allgemeine Formel dieser Komplexverbindung soll mit $MX_nL_xY_m$ charakterisiert werden. Mehrkernige Komplexverbindungen sowie Komplexverbindungen, welche erst in Kombination mit Ko-Katalysatoren als Hydrierungskatalysatoren fungieren, sowie geladene Komplexe (z.B. $Co(CN)_5^{3-}$) werden nicht in die Betrachtung einbezogen. Weiterhin werden Komplexverbindungen, welche erst bei hohen H_2-Drucken ($p_{H_2} > 5$ atü) und Temperaturen über 100 °C katalytische Eigenschaften aufweisen, nicht behandelt, da angenommen werden muß, daß der Katalysator nicht die Komplexverbindung, sondern ein Umwandlungs- oder Zersetzungsprodukt ist. Tabelle 1 zeigt Komplexverbindungen, für welche nachgewiesen wurde, daß sie homogene Hydrierungskatalysatoren sind (+). Weiterhin besagt (–), daß unter milden Bedingungen keine Hydrierung beobachtet wurde. Leere Felder bedeuten, daß die Verbindungen nicht auf katalytische Eigenschaften überprüft wurden.

III. Potentielle Katalysatoren aus der 8. Gruppe

Aus Tabelle 1 ergibt sich eindeutig, daß die meisten Hydrierungskatalysatoren Komplexverbindungen sind, welche Co, Rh, oder Ir als Zentralatom besitzen, wenn man nur jene Komplexe diskutiert, die bereits bei milden Reaktionsbe-

Tabelle 1. *Typen der Komplexverbindungen der 8. Gruppe, die bei milden Bedingungen als Hydrierungskatalysatoren fungieren. In den Literaturzitaten sind die weiteren Literaturhinweise enthalten*

Typ	Fe	Ru	Os	Co	Rh	Ir	Ni	Pd	Pt
ML_2X_2				+ 3)			_ 1)	_ 20)	_ 2)
ML_3X					+ 4),5)	+ 6),47)			
ML_2XCO					+ 7),8)	+ 7),9),10)			
$NOML_3$					+ 11)				
HML_3X		+ 12)							
HML_3CO				_ 13)	+ 14),15)	+ 15),16)			
ML_4X_2		+ 17)							
ML_3X_3					+ 18)				
HML_3XCO			+ 16)						
H_3ML_3				+ 13)		+ 19)			

dingungen hydrieren. Unter schärferen Reaktionsbedingungen werden auch Komplexverbindungen anderer Metalle brauchbare Hydrierungskatalysatoren. So können mit den Komplexen NiL_2X_2 [1], PdL_2X_2 [20] und PtL_2X_2 [2] bei 90 °C und p_{H_2} = 39,1 atü Hydrierungen ausgeführt werden. Das gleiche gilt für den Komplex $HCo(C\emptyset_3)_3CO$, welcher bei 150 °C und p_{H_2} = 50 atü Olefine hydriert [13]. Da auch für die aktiven Katalysatoren mit M = Rh und Ir generell gezeigt werden konnte, daß Temperaturerhöhung und höherer H_2-Druck die Hydrierungsgeschwindigkeit stark vergrößern, kann man als Arbeitshypothese annehmen, daß die meisten Komplexverbindungen der Tabelle 1 potentielle Katalysatoren sind, wenn die geeigneten Reaktionsbedingungen gefunden werden. Wie in späteren Abschnitten gezeigt wird, hängt die Aktivität solcher Hydrierungskatalysatoren außer von der Reaktionstemperatur, dem p_{H_2}-Druck und dem Lösungsmittel vor allem vom Zentralatom und der geeigneten Wahl der Liganden ab.

IV. Prinzipielle Hydrierungsmechanismen

Die Frage, wie die Aktivität eines homogenen Katalysators vom Zentralatom und den Liganden abhängt, kann auf zwei Wegen untersucht werden. Einerseits geben detaillierte Untersuchungen zum Reaktionsmechanismus die Möglichkeit, einen Einblick zu gewinnen, wie die Katalyse als chemische Reaktion abläuft. Mit dieser Information erhält man Anhaltspunkte, wie man in den Mechanismus eingreifen kann, um aktivere und selektivere Katalysatoren zu erhalten. Andererseits kann durch ein breites Spektrum von Modelluntersuchungen phänomenologisch geprüft werden, wie sich die Aktivität der Katalysatoren als Funktion der experimentell variierten Parameter ändert. Aus der Literatur ist zu erkennen, daß in den letzten Jahren beide Wege beschritten wurden.

Bezüglich des Reaktionsmechanismus der homogenen *katalytischen* Hydrierung gibt es zunächst drei prinzipielle Möglichkeiten [21]:

a) Der Katalysator Ka reagiert mit dem Wasserstoff nach Schema (1) zum Hydridkomplex KaH_2, der dann den Wasserstoff auf das Substrat S (ungesättigte Verbindung) überträgt. Dieser Mechanismus wird auch als *„hydride route"* bezeichnet [27].

$$Ka + H_2 \rightleftharpoons KaH_2 \xrightarrow{\ S\ } SH_2 + Ka \qquad (1)$$

b) Der Katalysator reagiert mit dem Substrat S nach (2) unter Bildung einer Komplexverbindung KaS, welche dann den Wasserstoff aufnimmt und überträgt (*„unsaturate route"*).

$$Ka + S \rightleftharpoons KaS \xrightarrow{\ H_2\ } KaSH_2 \longrightarrow SH_2 + Ka \qquad (2)$$

c) Der Katalysator reagiert sowohl mit Wasserstoff unter Bildung von KaH_2 und dem Substrat S unter Bildung von KaS. Die Hydrierung verläuft dann nach Schema (3). In diesem Reaktionsschema wird jeder Reaktant an einem verschiedenen Katalysatormolekül aktiviert.

$$Ka + H_2 \rightleftharpoons KaH_2$$
$$\longrightarrow SH_2 + 2\,Ka \qquad (3)$$
$$Ka + S \rightleftharpoons KaS$$

V. Quantitative Untersuchungen zum Reaktionsmechanismus

a) Hydrierung des Substrates S

In der Tabelle 2 sind Systeme zusammengestellt, für welche die Reaktionsgeschwindigkeit r der Hydrierung und die kinetischen Ansätze, die sich aus den experimentellen Daten ableiten lassen, quantitativ untersucht wurden. Stets wurde die Reaktionsgeschwindigkeit aus der Abnahme des Wasserstoffes nach Gleichung I bestimmt.

Tabelle 2. *Systeme homogener, katalytischer Hydrierung, welche quantitativ untersucht wurden*

Nr.	KaL_x	Substrat S	Lös.-Mittel	Temp °C	$f[S]_0$	$f[KaL_x]_0$	$f(t)$	$f(p_{H_2})$	$f[L]_0$	$f(\text{solv})$	Lit.
	$RhXL_3$	1 Hepten	Benzol	25	+						22)
	X = Cl	1 Hexin	Benzol	25	+						
1	L = PΦ_3	Cyclohexen	Benzol	25	+	+	+	+			
	X = Cl, Br, J	Cyclohexen	Benzol	25	+						
	L = PΦ_3										
	X = Cl; L = PΦ_3	Cyclohexen	diverse 1:1	25	+					+	
2	X = Cl; L = PΦ_3	diverse	C_6H_6/C_6H_{12}	25	+						23)
3	$RhHCO(PΦ_3)_3$	1-Hexen	Benzol	25	+	+	+	+	+		24)
		1-Decen	Benzol	25	+	+	+	+			
4		1-Hepten	Toluol	25	+	+			+		25)
		MDME	Toluol	25	+	+			+		
5	$IrX(CO)L_2$	1-Hepten	Toluol	80	+	+			+		
	X = Cl, Br, J, SCN	MDME	Toluol	80	+	+	+		+		26)
	L = $P(C_6H_{11})_3$; PΦ_3; $P(iC_3H_7)_3$; $P(OΦ)_3$	MA	Toluol	80	+	+			+		50)
6	X = Cl, Br, J	MA	$HCON(CH_3)_2$	80	+	+		+			9)
	L = PΦ_3										

75

Tabelle 3. *Vorgeschlagene Reaktionsmechanismen und kinetische Ansätze zu den untersuchten Systemen der Tabelle 2 (K = Gleichgewichtskonstante, k = Geschwindigkeitskonstante)*

Nr. der Tab. 2	Katalysator	Mechanismen	Schema	kinetischer Ansatz	Gleichung	Lit
1	$RhXL_3$	$KaL_3 \rightleftharpoons KaL_2 + L$ ($\alpha_{Diss} \sim 100\%$) $KaL_2 \xrightarrow{+H_2, K_1} KaL_2H_2$ $KaL_2 \xrightarrow{+S, K_2} KaL_2S$ $KaL_2S \xrightarrow{+S, k'} KaL_2 + SH_2$ $KaL_2S \xrightarrow{+H_2, k''} KaL_2 + SH_2$	(4)	$r = \dfrac{\{k'·K_1 + k''·K_2\}·p·[S]_0·[KaL_3]_0}{1 + K_1·p + K_2·[S]_0}$		22)
2				mit $k'' \sim 0$ $r = \dfrac{k'·K_1·p[S]_0·[KaL_3]_0}{1 + K_1·p + K_2·[S]_0}$	II	23)
3	$RhH(CO)L_3$	$KaL_3 \rightleftharpoons KaL_2 + L$ ($\alpha_{Diss} \sim 100\%$) $KaL_2 \xrightarrow{+S, K_2} KaL_2S \xrightarrow{+H_2, k_1} KaL_2SH_2 \rightarrow KaL_2 + SH_2$	(5)	$r = \dfrac{k_1·K_2[H_2]·[S]_0·[KaL_3]_0}{1 + K_2[S]_0}$		24)
4				mit $1/K_2 = K$ $r = \dfrac{k_1·[H_2]·[S]_0·[KaL_3]_0}{K + [S]_0}$	III	25)
5	$IrX(CO)L_2$	$KaL_2H_2 \underset{k_1}{\xleftrightarrow{-H_2}} KaL_2 \underset{K_2}{\xleftrightarrow{+S}} KaL_2S \underset{K_3}{\xleftrightarrow{-L}}$ $KaLS \xrightarrow[k_4]{+H_2} KaLSH_2 \longrightarrow KaL + SH_2$	(6)	$r = \dfrac{k_4·K_3[S]_0[H_2][KaL_2]_0}{\{K_3[S]_0 + [L]\}\{K_2 + K_2[H_2]/K_1\}}$	IV	26)
6		$KaL_2H_2 \underset{K_H}{\xleftrightarrow{-H_2}} KaL_2 \underset{k_{-1}}{\overset{k_1}{\xleftrightarrow{}}} KaL + L$ $KaL \underset{K_m}{\xleftrightarrow{+S}} KaLS \xrightarrow[k_2]{+H_2} KaL + SH_2$	(7)	$r = \dfrac{k_2·K_m[H_2]·[KaL_2]_0·[S]_0}{1 + K_m[S]_0 + [L]\{1 + K_H·[H_2]\}/K}$	V	9)

$$r = -\delta H_2/\delta t = -\delta [S]/\delta t \qquad\qquad \text{I}$$

In der Tabelle 2 bedeuten $r = f(p_{H_2})$; $r = f[S]_0$; $r = f[KaL_x]_0$; $r = f(t)$; $r = f[L]_0$
und $r = f(\text{solv})$, daß die Reaktionsgeschwindigkeit r
als Funktion des Wasserstoffdruckes,
der Konzentration des Katalysators,
der Temperatur,
der Konzentration an freiem Liganden L und
des Lösungsmittels
untersucht wurden. Da, wie in der Diskussion gezeigt wird, die Dissoziation der
ungeladenen Liganden L des Katalysators eine bedeutende Rolle spielt, soll die
als Katalysator eingesetzte Komplexverbindung mit KaL_x abgekürzt werden,
z.B. $RhX(CO)L_2 \equiv KaL_2$. Als weitere Abkürzungen werden verwendet:
MDME = Maleinsäuredimethylester,
MA   = Maleinsäureanhydrid,
FDME = Fumarsäuredimethylester,
ACAE = Acrylsäureäthylester.
Die Reaktionsmechanismen und die daraus abgeleiteten kinetischen Ansätze
sind in Tabelle 3 zusammengestellt. Der Index 0 z.B. $[S]_0$ bedeutet, daß es sich
um die Einwaage- und nicht die Gleichgewichts-Konzentration des Substrates
handelt; $[H_2]$ ist die Konzentration des im Reaktionsmedium gelösten Wasser-
stoffes, $p \equiv$ Druck des H_2 über dem Reaktionsmedium.

b) Hydrierung des Katalysators

Aus Tab. 3 Nr. 1 erkennt man, daß bei der ,,hydride route" die Hydrierung des
Katalysators KaL_x eine wesentliche Rolle spielt. Falls sich das Gleichgewicht nach
Schema (8) nicht sehr schnell einstellt

$$KaL_x + H_2 \overset{K_1}{\rightleftharpoons} KaL_xH_2 \qquad\qquad (8)$$

kann die Kinetik der Hydrierung des Substrates im Falle der ,,hydride route"
durch die Reaktionsgeschwindigkeit r_{H_2} der Bildung von KaL_xH_2 kontrolliert
werden, wobei r_{H_2} nach Gleichung VI definiert ist.

$$r_{H_2} = -\delta H_2/\delta t = +\delta[KaL_xH_2]/\delta t \qquad\qquad \text{VI}$$

Quantitative Bestimmungen von r_{H_2} wurden bisher nur in einigen Fällen (s.
Tabelle 4) vorgenommen, obwohl der Kenntnis der r_{H_2}-Werte bei der Entschei-
dung über die möglichen Reaktionsmechanismen eine große Bedeutung zukommt.

VI. Diskussion der Reaktionsmechanismen

Wie Tabelle 2 zeigt, wurde als Mechanismus der homogenen katalytischen Hydrierung
mit $RhH(CO)(P\emptyset_3)_3$ und $IrX(CO)L_2$ die ,,unsaturate route" wahrscheinlich ge-

macht. Der in [9] vorgeschlagene Mechanismus ist mit dem der [26] praktisch identisch, da die Frage, ob erst das Substrat S an $IrX(CO)L_2$ anlagert und dann der Ligand L abgespalten wird [26] oder die Reihenfolge umgekehrt ist [9], sich experimentell nicht exakt beantworten läßt. Nachdem sich jedoch einige Komplexe wie $IrX(CO)[P(O\emptyset)_3]_2$ (MDME) isolieren lassen [49], wurde der Mechanismus nach Schema (6) (in Tabelle 2) vorgeschlagen.

Für $RhXL_3$ wurden zunächst die *hydride* und *unsaturate route* erörtert [22]. Man glaubt jedoch auf Grund experimenteller Befunde schließen zu müssen [22], daß $k'' = \sim 0$ ist (Tabelle 3, Gl. II), da Äthylen, das mit $RhCl(P\emptyset_3)_3$ den isolierbaren Komplex $RhCl(P\emptyset_3)_2C_2H_4$ bildet, nicht hydriert wird. Spätere Untersuchungen zeigten aber, daß C_2H_4 von $RhCl(P\emptyset_3)_3$ hydriert wird und daß auch der Komplex $RhCl(P\emptyset_3)_2C_2H_4$ selbst ein Hydrierungskatalysator [27] ist, so daß für $RhXL_3$ die *unsaturate route* nicht ausgeschlossen werden darf [27]. Nun kann aber für den Fall, daß die Reaktionsgeschwindigkeit r der Hydrierung des Substrates wesentlich größer als die Reaktionsgeschwindigkeit r_{H_2} der Hydrierung des Katalysators ist, die *hydride route* ausgeschlossen werden, da in der *hydride route* die Hydrierung des Substrates nie schneller sein kann, als die Hydrierung des Katalysators [28]. In der Tabelle 4 sind nun diesbezügliche Werte von r und r_{H_2} gegenübergestellt. Die r_{H_2}-Werte sind auf die Zeit $t = 0$ extrapoliert, während die r-Werte die Mittelwerte im Anfang der Hydrierung wiedergeben.

Tabelle 4. *Vergleich korrespondierender r_{H_2} und r-Werte (Lösungsmittel: Toluol)*

Nr.	KaL_x	$[KaL_x]_0$ $mMol \cdot l^{-1}$	Temp. °C	Lit.	r_{H_2} $(mMol \cdot l^{-1}$ $min^{-1})$	r $(mMol \cdot l^{-1}$ $min^{-1})$	S	[S] $Mol \cdot l^{-1}$	Lit.
1	$RhCl(P\emptyset_3)_3$	2,0	25	[28]	1,7	8,2	MDME	0,73	[29]
2	$RhCl(P\emptyset_3)_3$	2,0	25	[28]	1,7	5,4	Cyclohepten	0,73	[29]
3	$RhBr(P\emptyset_3)_3$	2,0	25	[28]	2,3	5,6	Cyclohepten	0,73	[29]
4	$IrCl(CO)[P(C_6H_{11})_3]_2$	2,0	80	[30]	0,0053	19,7	MA	0,8	[31]
5	$IrCl(CO)[P(C_6H_{11})_3]_2$	2,0	80	[30]	0,0053	1,0	MDME	0,8	[31]
6	$IrCl(CO)[P(iC_3H_7)_3]_2$	2,0	80	[30]	0,035	1,06	MA	0,8	[31]
7	$IrCl(CO)(P\emptyset_3)_2$	2,0	80	[30]	3,1	8,9	1-Hepten	0,8	[31]
8	$IrCl(CO)(P\emptyset_3)_2$	2,0	80	[30]	3,1	10	ACAE	0,8	[31]
9	$IrCl(CO)(P\emptyset_3)_2$	2,0	80	[30]	3,1	6,6	Styrol	0,8	[31]

Wie die Tabelle 4 zeigt, sind für die Systeme Nr. 1–9 die Hydrierungsgeschwindigkeiten r für die Substrate S größer und z.T. sogar um Größenordnungen größer als die r_{H_2}-Werte für die Hydrierung des Katalysators. Wenn man weiterhin in Betracht zieht, daß sich die in Tabelle 4 angegebenen r_{H_2}-Werte auf die Einwaagekonzentrationen $[KaL_x]_0$ beziehen und während der Hydrierung immer der

Hydridkomplex KaL_xH_2 [22,29)] im Gleichgewicht vorhanden ist, so muß das Verhältnis von r_{H_2} zu r noch größer sein. Für die Beispiele der Tabelle 4 kann somit die *hydride route* nicht der geschwindigkeitsbestimmende Schritt bei der Hydrierung sein.

Für den Katalysator $RhH(CO)(P\emptyset)_3)_3$ (Tabelle 2, Nr. 3 und 4) wurde die *hydride route* nicht diskutiert, da die Bildung des Hydridkomplexes $RhH_3(CO)(P\emptyset_3)_3$

$$RhH(CO)P\emptyset_3)_3 + H_2 \rightleftharpoons RhH_3(CO)(P\emptyset_3)_3 \qquad (9)$$

nach dem Schema (9) *nicht beobachtet* wurde [24)].

Im Gegensatz dazu steht jedoch der Katalysator $IrX(CO)L_2$ mit seinem Hydridkomplex $IrX(CO)L_2H_2$ unter den Bedingungen der Hydrierung immer im Gleichgewicht. Die Beispiele Nr. 4–9 der Tabelle 4 zeigen jedoch, daß die r-Werte für die Hydrierung des Substrates S größer sind als die r_{H_2}-Werte für die Hydrierung des Katalysators, so daß auch mit diesem Katalysator die Hydrierung zum überwiegenden Teil über die *unsaturate route* verlaufen muß. Analog zu $RhXL_3$ konnte jedoch auch für $IrX(CO)L_2$ kein eindeutiges Experiment gefunden werden, das die *hydride route* völlig ausschließt.

VII. Probleme bei katalytischen Hydrierungsreaktionen

a) Allgemeine Bemerkungen

In Forschung und Technik erhebt sich stets die Frage, welcher Katalysator-Typ für die gestellte Aufgabe benutzt werden kann, wie groß seine Aktivität und seine Selektivität sind. Einen potentiellen Hydrierungskatalysator erhält man sofort aus Tabelle 1. Er wird jedoch noch keine optimale Reaktionsführung und Ausbeute geben, so daß Informationen darüber vorhanden sein müssen, wie seine Aktivität und Selektivität gezielt geändert werden können. Nützlich sind dann Modellvorstellungen und allgemeine Regeln. Es sollen daher die zahlreichen in der Literatur bereits vorliegenden halbquantitativen Untersuchungen zur homogenen katalytischen Hydrierung unter diesem Gesichtspunkt referiert und diskutiert werden.

b) Modellvorstellung zur homogenen Katalyse

Die Grundelemente der folgenden Modellvorstellung [32)] werden mehr oder minder in fast allen Arbeiten über homogene Katalyse zur Interpretation der Ergebnisse herangezogen. Sie sind in Abb. 1 bildlich dargestellt.
1. In einem homogenen Katalysator, der zum Typ der Komplexverbindungen gehört, ist das Zentralatom das Reaktionszentrum RZ.
2. Durch geeignete Wahl des Zentralatoms und der geladenen Liganden X und/ oder der ungeladenen Liganden L kann über die σ-Bindung und/oder π-Elek-

tronenrückgabebindung in der Bindung M–X bzw. $M \underset{\sigma}{\overset{\pi}{\rightleftharpoons}} L$ die Elektronendichte am Reaktionszentrum variiert werden.

3. Am Zentralatom muß mindestens eine unbesetzte Koordinationsstelle vorhanden sein.

4. Befindet sich an der unbesetzten Koordinationsstelle ein „leeres" Reaktionsorbital RO (Unterschuß der Ladungsdichte), so handelt es sich um einen Acceptor-Katalysator Ka_A (Abb. 1a). Ist das Reaktionsorbital gefüllt (Überschuß an Ladungsdichte), so hat man einen Donator-Katalysator Ka_D (Abb. 1b).

5. Die Aktivierung des Substrates S geschieht durch Anlagerung an die unbesetzte Koordinationsstelle.

6. Die Anlagerung des Substrates S an das Reaktionszentrum darf nicht irreversibel sein, damit das Substrat weiter reagieren kann.

7. Damit Bedingung 6 erfüllt wird, muß durch das Zentralatom und die Liganden X und L im Reaktionsorbital RO eine optimale Elektronendichte ED einjustiert werden.

8. Besitzt der „Katalysator" keine unbesetzte Koordinationsstelle, so muß die Elektronendichte am Reaktionszentrum so einjustiert sein, daß er in Lösung entweder einen Liganden abdissoziieren kann, oder durch Anlagerung des Substrates seine Koordinationszahl erhöhen kann.

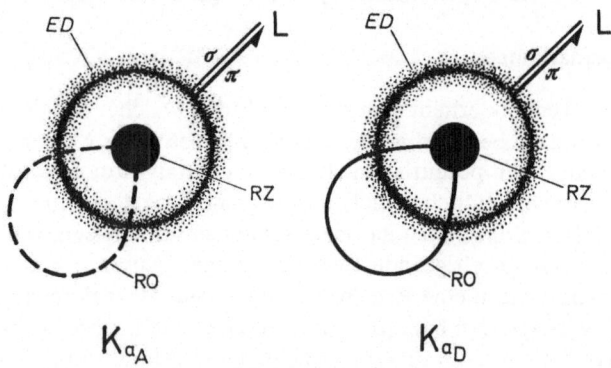

Abb. 1. Zum Katalysator-Modell

Aus dieser zweifelsohne sehr vereinfachten Modellvorstellung erhebt sich die Frage, welche Eigenschaften des Zentralatoms und der Liganden als Maß für ihre Änderung der Elektronendichte am Reaktionszentrum genommen werden können. Für die Zentralatome bietet sich hierfür ihre formale Ladung bzw. Wertigkeit an und dann ihr kovalenter Radius als Funktion der gesamten Elektronen. Für die geladenen Liganden X kann ihre Elektronegativität genommen werden und für die ungeladenen Liganden L ihre π-Acceptorstärke, deren relative Abstufung für die hier interessierenden Verbindungen bekannt ist [33].

c) Dissoziationskonstante K_2 für den Substratkomplex des Katalysators als Funktion von X, L, S und M

Nachdem die quantitativen Untersuchungen zum Reaktionsmechanismus gezeigt haben, daß die Katalyse der Hydrierung zum überwiegenden Teil über die *unsaturate route* verläuft, war es von Interesse, wie die Dissoziationskonstante K_2 nach Schema (10)

$$KaL_x + S \; \underset{K_2}{\rightleftharpoons} \; KaL_xS \qquad (10)$$

von der Variation der Elektronendichte am Reaktionszentrum abhängt. Quantitative Untersuchungen zu diesem Problem wurden mit Vaska's Hydrierungskatalysator $IrX(CO)L_2$ durchgeführt [34]. In der Tabelle 5 ist die Änderung der Dissoziationskonstanten K_2 als Funktion von L und S und in Tabelle 6 als Funktion von X und S zusammengestellt.

Tabelle 5. *Dissoziationskonstante K_2 für $IrCl(CO)L_2S \rightleftharpoons IrCl(CO)L_2 + S$ bei t = 20 °C in Toluol als Funktion der Bindung Ir $\rightleftharpoons$ L und der π-Acceptorstärke von S* [34]

S	$K_2 \cdot 10^4 (Mol \cdot 1^{-1})$		
	Ir $\rightleftharpoons$ P(O$\emptyset_3$)$_3$	Ir $\rightleftharpoons$ P$\emptyset_3$	Ir $\rightleftharpoons$ P(C$_6$H$_{11}$)$_3$
Maleinsäureanhydrid	1,1	28	2.200
Maleinsäuredimethylester	10	2200	$> 3 \cdot 10^{5*}$
Acrylsäureäthylester	260	35000	$> 3 \cdot 10^5$
Fumarsäuredimethylester	2200	$> 3 \cdot 10^5$	$> 3 \cdot 10^5$
1-Hepten	94000	$> 3 \cdot 10^5$	$> 3 \cdot 10^5$

(Abnahme der π-Acceptorstärke)

* $> 3 \cdot 10^5$ bedeutet: K_2 spektroskopisch nicht meßbar.

Aus den Zeilen der Tabelle 5 erkennt man, daß mit abnehmender π-Elektronenrückgabebindung in Ir $\rightleftharpoons$ L, also abnehmender π-Acceptorstärke von L, die Dissoziationskonstante K_2 zunimmt. Dieser Effekt ist unabhängig von den in der Tabelle 5 angegebenen Substraten S.

Betrachtet man die Spalten der Tabelle 5, so ergibt sich weiterhin, daß auch mit abnehmender π-Acceptorstärke des Substrates K_2 zunimmt. Dieser Effekt ist wieder unabhängig von der Ir $\rightleftharpoons$ L Bindung. Beide Effekte liegen auch bei den Bromo- und Jodo-Komplexen $IrBr(CO)L_2$ und $IrJ(CO)L_2$ in der gleichen Richtung und Größenordnung.

Tabelle 6. *Dissoziationskonstante K_2 für $IrX(CO)[PO\varphi_3]_2S \rightleftharpoons IrX(CO)[P(O\varphi_3]_2 + S$ bei $t = 20\,°C$ in Toluol als Funktion der Bindung Ir–X und der π-Acceptorstärke des Substrates S*

S			$K_2 \cdot 10^4 (\text{Mol} \cdot 1^{-1})$	
		Ir◄─Cl	Ir◄─Br	Ir ── J
Abnahme der π-Acceptorstärke	Maleinsäureanhydrid	1,1	2,2	4,6
	Maleinsäuredimethylester	10	23	95
	Arcrylsäureäthylester	260	450	950
	Fumarsäuredimethylester	2200	7500	$> 3 \cdot 10^5$
	1-Hepten	94000	$> 3 \cdot 10^{-5}$	$> 3 \cdot 10^5$

Tabelle 6 zeigt analog zu Tabelle 5, daß die Dissoziationskonstante K_2 auch in eindeutiger Weise von der Bindung Ir–X und der π-Acceptorstärke des Substrates S abhängt. Die K_2-Werte nehmen unabhängig vom Substrat in der Reihe für X = Cl < Br < J zu (Zeilen der Tabelle 6) und bei gleichem X nimmt K_2 mit abnehmender π-Acceptorstärke des Substrates zu [34].

Über den Einfluß des Zentralatoms M des Katalysators auf die Komplexbildung liegen keine systematischen und quantitativen Messungen vor. Die Ergebnisse qualitativer Untersuchungen sind in der Tabelle 7 zusammengestellt.

Tabelle 7. *Vergleich der Komplexbildungstendenz von $MX(CO)L_2$*

Komplex	S	Bemerkung	Lit.
$IrX(CO)(P\varphi_3)_2$	$C_2(CN)_4$	IrL_2S stabiler	[35]
$RhX(CO)(P\varphi_3)_2$	$C_2(CN)_4$	als RhL_2S	[35]
$IrCl(CO)(P\varphi_3)_2$	C_2H_4	Komplexbildung	[7]
$RhCl(CO)(P\varphi_3)_2$	C_2H_4	keine Komplexbildung	[35]

Man erkennt, daß der Iridium-Komplex die größere Tendenz besitzt, mit dem Substrat S eine Komplexverbindung zu bilden. Die Komplexe $IrX(CO)L_2S$ bilden sich sofort bei Zugabe der Substratlösung zum gelösten Katalysator $IrX(CO)L_2$.

d) Geschwindigkeit der Hydrierung des Katalysators als Funktion der Liganden X und L

Im Gegensatz zum Substrat ist die Reaktion des Wasserstoffes mit dem Katalysator nach Schema (11) eine Zeitreaktion, die nur für einige Katalysatoren quantitativ und systematisch als Funktion X und L untersucht wurde.

Aus der Tabelle 8 ersieht man wieder, daß für den Katalysator $IrX(CO)L_2$ ein

$$KaL_x + H_2 \underset{}{\overset{k_1}{\rightleftharpoons}} KaL_xH_2 \qquad (11)$$

eindeutiger Zusammenhang zwischen der Geschwindigkeitskonstanten k_1 der Hydridbildung und seinen Liganden X und L besteht [36]. Bei festgehaltenem Liganden X nimmt k_1 mit zunehmender π-Acceptorstärke des Liganden L zu, erreicht bei L = P(p-tolyl)$_3$ seinen maximalen Wert und fällt dann wieder ab.

Auch der Einfluß von X auf k_1 ist eindeutig. Bei festgehaltenen Liganden L nimmt k_1 in allen Fällen in der Reihe für X = Cl < Br < J zu. Der gleiche Effekt wurde für $IrX(CO)(P\emptyset_3)_3$ gelöst in Benzol gefunden [37].

Tabelle 8. *Geschwindigkeitskonstanten k_1 für die Reaktion*

$$IrX(CO)L_2 + H_2 \underset{}{\overset{k_1}{\rightleftharpoons}} IrX(CO)L_2H_2$$

als Funktion der Liganden X und L in Toluol [36]

X	IrX(CO)L$_2$ L	$k_1 \cdot 10^2$ $1 \cdot Mol^{-1} \cdot sec^{-1}$	Temp. °C
	P(C$_6$H$_{11}$)$_3$	1,6	80
	P(iC$_3$H$_7$)$_3$	10,4	80
	P(C$_4$H$_9$)$_3$	525	80
Cl	P(CH$_2\emptyset$)$_3$	767	80
	P(p-tolyl)$_3$	1150	80
	P($\emptyset_3$)	838	80
	P(O$\emptyset_3$)$_3$	121	80
Cl		0,35	50
Br	P(C$_6$H$_{11}$)$_3$	6,5	50
J		224	50
Cl		2,5	50
Br	P(iC$_3$H$_7$)$_3$	33	50
J		108	50
Cl		31	20
Br	P$\emptyset_3$	580	20
J		>4000	20
Cl		3,6	20
Br	P(O$\emptyset$)$_3$	192	20
J		~5600	20

(In der Spalte L vertikal: *Zunahme der π-Acceptor-Stärke*)

e) Dissoziation der Komplexe $MX_nL_xY_m$ als Funktion ihrer Liganden X und L, des Zentralatoms M und des Lösungsmittels

Nach Punkt 8 der Modellvorstellung zur Katalyse muß die Elektronendichte am Zentralatom so einjustiert werden, daß durch Anlagerung des Substrates die Koordinationszahl erhöht werden kann oder durch Abdissoziation eines Liganden eine unbesetzte Koordinationsstelle gebildet wird, an welche das Substrat S angelagert wird. In der Tabelle 9 sind die Katalysatoren zusammengestellt, für welche osmometrische Molekulargewichtsbestimmungen durchgeführt wurden. Die Messungen entbehren einer strengen Systematik, zeigen jedoch die Unterschiede zwischen vergleichbaren Komplexen.

Tabelle 9. *Dissoziation der Liganden L von Komplexverbindungen, die Hydrierungskatalysatoren sind, nach $KaL_x \rightleftharpoons KaL_{x-1} + L$ (* in sehr verdünnten Lösungen)*

Nr.	Katalysator	Lösungsmittel	α_{Diss}	Lit.
1	$IrX(CO)L_2$	Toluol	~ 0	38)
2	$RhCl(CO)(P\varphi_3)_2$	Toluol	$< 0,05$	38)
3	$RhBr(CO)(P\varphi_3)_2$	Toluol	$\sim 0,15$	38)
4	$RhJ(CO)(P\varphi_3)_2$	Toluol	$\sim 0,20$	38)
5	$IrCl(P\varphi_3)_3$	Benzol	$\sim 0,23$	6)
6	$IrCl(P\varphi_3)_3$	$CHCl_3$	$\sim 0,35$	6)
7	$RhCl(P\varphi_3)_3$	Benzol	$\sim 0,95$	22)
8	$RhCl(As\varphi_3)_3$	Benzol	$\sim 0,55$	40)
9	$RhCl(As\varphi_3)_3$	$CHCl_3$	$\sim 0,93$	40)
10	$RhCl(Sb\varphi_3)_3$	$CHCl_3$	$\sim 0,93$	40)
11	$IrH(CO)(P\varphi_3)_3$	Benzol	$< 0,02$	39)
12	$RhH(CO)(P\varphi_3)_3$	Benzol	$\sim 1,0$	39)
13	$RhH(CO)(P\varphi_3)_3$	Benzol	$\sim 1,3*$	41)
14	$RhH(CO)(P\varphi_3)_3$	Cyclohexan	$\sim 0,28$	41)
15	$RhCl(P\varphi_3)_3$	Toluol	~ 1	38)
16	$RhBr(P\varphi_3)_3$	Toluol	~ 2	38)
17	$RhJ(P\varphi_3)_3$	Toluol	$\sim 1,35$	38)
18	$RhCl(P(O\varphi)_3)_3$	Toluol	~ 0	38)
19	$RhBr(P(O\varphi)_3)_3$	Toluol	$\sim 0,16$	38)
20	$RhJ(P(O\varphi)_3)_3$	Toluol	$\sim 0,6$	38)

Der Einfluß des Zentralatoms auf die Dissoziation der Liganden ist eindeutig. Bei den Komplexen $MX(CO)_2L_2$ (Nr. 1–4), $MClL_3$ (Nr. 5–7), und $MH(CO)L_3$ (Nr. 11–14) ist immer der Rhodiumkomplex wesentlich stärker dissoziiert als der Iridiumkomplex.

Der Einfluß des geladenen Liganden X bewirkt eine stärkere Dissoziation in der Reihe α_{Diss} für X = Cl < Br < J (Nr. 2–4, 15–17, 18–20).

Bezüglich der ungeladenen Liganden L zeichnet sich ab, daß Komplexe mit L = $P\varnothing_3$ stärker dissoziiert sind, als mit L = $P(O\varnothing)_3$ (Nr. 15 und 18; 16 und 19; 17 und 20).

Diese Beobachtungen stehen im Einklang mit Ligandenaustauschversuchen [42] an den Komplexen $IrX(CO)L_2$ und $RhX(CO)L_2$ nach (12) und (13). Die Rh-Komplexe tauschen die Liganden

$$MX(CO)L_2 \;+\; L' \;\rightleftharpoons\; MX(CO)LL_2' \;+\; L \qquad (12)$$

$$MX(CO)LL' \;+\; L' \;\rightleftharpoons\; MX(CO)L_2' \;+\; L \qquad (13)$$

schneller aus als die Ir-Komplexe und bezüglich der Liganden L ist der Austausch für L = $P\varnothing_3 > P(O\varnothing)_3 \gg P(C_6H_{11})_3$.

Besondere Bedeutung kommt den Komplexen Nr. 13, 16 und 17 der Tabelle 9 zu, für welche ein Dissoziationsgrad $\alpha > 1$ nachgewiesen wurde, da diese Komplexe in sehr verdünnten Lösungen nach Schema (14)

$$KaL_x \;\rightleftharpoons\; KaL_{x-1} + L \;\rightleftharpoons\; KaL_{x-2} + 2\,L \qquad (14)$$

einen 2. Liganden abspalten und somit zwei unbesetzte Koordinationsstellen bilden, was, wie später gezeigt wird, ihre katalytische Reaktivität stark erhöht.

Auch der Einfluß des *Lösungsmittels* ist aus Tabelle 9 zu erkennen. So ist die Dissoziation in dem inerten Lösungsmittel Cyclohexan kleiner als in Benzol (Nr. 14, 12 und 13) und in Benzol kleiner als in Chloroform (Nr. 5 und 6, 8 und 9). Die Fähigkeit des $CHCl_3$ die Dissoziation der Liganden zu begünstigen ist deswegen von Bedeutung, da es in einigen Fällen gelang, von Katalysatoren, bei welchen in Lösung ein Ligand abdissoziiert, ihre Solvatkomplexe KaL_{x-1} (solv) mit solv $\equiv CH_2Cl_2$ zu isolieren. Einige Beispiele sind in der Tabelle 10 enthalten.

Tabelle 10. *Isolierte Solvatkomplexe von Katalysatoren*

Nr.	Komplexe	Lit.
1	$RhCl(P\varnothing_3)_2H_2(CH_2Cl_2)_{0,5}$	[22]
2	$RhCl(P\varnothing_3)_2H_2(DMFA)_{0,5}$	[22]
3	$RhCl(P\varnothing_3)_2H_2(CH_3CO_2\,et)_{0,5}$	[22]
4	$RhCl(As\varnothing_3)_2H_2(CH_2Cl_2)_{0,5}$	[40]
5	$[Rh(CO)(P\varnothing_3)_2(CH_2Cl_2)]_2$	[41]
6	$[Rh(CO)(P\varnothing_3)_2(C_2H_5OH)]_2$	[41]
7	$RhCl(P\varnothing_3)_2(CS_2)$	[44]
8	$IrJ(CO)(P\varnothing_3)_2(CS_2)$	[44]
9	$IrCl(CO)(P\varnothing_3)_2(CS_2)$	[44]

Es wurde bereits bei der Beobachtung, daß Katalysatoren in Lösung den Liganden L abspalten die Hypothese aufgestellt, daß der Komplex in Lösung nicht mit einer völlig unbesetzten Koordinationsstelle vorliegt. In Lösung ist diese, wenn auch nur sehr locker mit einem Solvatmolekül nach Schema (15) besetzt [43].

$$RhCl(P\emptyset_3)_3 \quad \overset{solv}{\rightleftharpoons} \quad RhCl(P\emptyset_3)_2 \text{ (solv)} + P\emptyset_3 \qquad (15)$$

Die Tatsache Solvatkomplexe isolieren zu können, bestätigt diese Hypothese. Von weiterem Interesse ist, daß auch Katalysatoren wie $IrX(CO)L_2$, die, ohne Verlust eines Liganden an ein vorhandenes, unbesetztes Orbital das Substrat S unter Erhöhung ihrer Koordinationszahl und Bildung von z.B. $IrX(CO)L_2S$ anlagern, auch mit Lösungsmitteln wie CS_2, entsprechende *Solvatkomplexe* bilden (Tabelle 10, Nr. 8 und 9).

Der Einfluß des Lösungsmittel auf die Aktivität eines Hydrierungskatalysators wird also unter dem Gesichtspunkt zu diskutieren sein, wie groß seine Fähigkeit ist, die Abdissoziation eines Liganden zu erhöhen und wie stark das Lösungsmittelmolekül an die freie Koordinationsstelle gebunden ist. Diese Bindung darf auf keinen Fall so groß sein, daß die Koordinationsstelle für die Anlagerung des Substrates S blockiert wird. So konnte gezeigt werden, daß die Hydrierungsgeschwindigkeit für den Katalysator $IrCl(CO)(P\emptyset_3)_2$ sehr stark abnimmt, wenn dem Lösungsmittel Toluol in geringen Mengen der gute Donator $N(C_2H_5)_3$ zugegeben wird und gar auf Null sinkt, wenn $[N(C_2H_5)_3] > 0,008$ Mol $\cdot$ 1^{-1} wird [45]. Analog wird die Hydrierungsgeschwindigkeit $r = 0$ für $RhX(P\emptyset_3)_3$, wenn dem Lösungsmittel stark koordinierende Verbindungen wie $P\emptyset_3$, Thiophen, 8-Hydroxychinolin, Pyridin oder Acetonitril hinzugefügt werden [22].

Lösungsmitteleinflüsse, welche zu einer chemischen Umsetzung mit dem Katalysator führen, sollen hier nicht diskutiert werden, obwohl dadurch die Katalysatoraktivität erhöht werden kann. So hydriert $RuCl_2(P\emptyset_3)_2$ in C_6H_6 nur schlecht, aber gut im Mischlösungsmittel C_2H_5OH/C_6H_6, da sich durch Reaktion mit C_2H_5OH der sehr gute Katalysator $RuHCl(P\emptyset_3)_3$ bildet [17]. In der Praxis wird man natürlich gleich diesen Katalysator einsetzen oder in situ herstellen.

f) Kriterium für die Katalysatoraktivität

Die unter den Punkten c – e) dargelegten experimentellen Befunde zeigten Gesetzmäßigkeiten für die Reaktionen der Katalysatoren mit H_2, dem Substrat S und dem Lösungsmittel als Funktion der Liganden X und L und des Zentralatoms. Mit diesen Informationen soll nun versucht werden, ob ein Zusammenhang zwischen der Aktivität eines Hydrierungskatalysators bezüglich eines Substrates S und der Variation der Liganden X und L und des Zentralatoms gezeigt werden kann, da eine gezielte Variation dieser Parameter letztlich das zentrale Problem jeder Katalyse ist. Als quantitatives Maß für die Aktivität eines Katalysators muß man die Geschwindigkeitskonstante für die betrachtete Reaktion nehmen. Unglücklicherweise setzt dies aber voraus, daß auch der Reaktionsmechanismus exakt bekannt ist. Abgesehen von der Tatsache, daß das Erarbeiten eines Reaktionsmechanismuses äußerst zeitraubend ist, kann die Aussage kinetischer Messungen nie definitiv sein, da durch jedes gezielte kinetische Experiment die Zahl der möglichen Mechanismen nur verkleinert werden kann. Weiterhin zeigen bereits die kinetischen Ansätze der Tabelle 3, daß selbst für einfache Hydrierungsreaktionen zur Berechnung der k-Werte mehrere Gleichgewichtskonstanten explizit bekannt sein müssen, und außerdem noch die Reaktionsgeschwindigkeit r. Da in der Katalyse, von der Anwendung her gesehen, immer der Umsatz pro Zeiteinheit, also die Reaktionsgeschwindigkeit r die wichtigste Kenngröße ist, soll als relatives Maß für die Aktivität eines Katalysators seine Reaktionsgeschwindigkeit r unter vergleichbaren Bedingungen genommen werden. Zweifelsohne gehen dadurch mechanistische Informationen verloren, aber die Information wie sich r als Funktion des Katalysators, des Substrates und des Lösungsmittels ändert, ist in jedem interessierenden Falle immer und in sehr kurzer Zeit zu erhalten. Im folgenden wird daher die Reaktionsgeschwindigkeit r zu Beginn der Katalyse, um Einflüsse einer eventuellen chemischen Veränderung des Katalysators auszuschließen, als relatives Maß seiner Aktivität genommen.

g) r als Funktion der Liganden X

Für den Katalysator $IrX(CO)L_2$ wurde die Reaktionsgeschwindigkeit r als Funktion der Liganden X unter streng vergleichbaren Bedingungen untersucht [10] Tabelle 11 zeigt, daß unabhängig vom Liganden L und unabhängig vom Substrat S die Reaktionsgeschwindigkeit r immer für X = Cl > Br > J ist.
Die Ergebnisse der analogen Untersuchungen mit $Rh(CO)L_2$ [8] sollen hier nicht diskutiert werden, da nicht $RhX(CO)L_2$, sondern der in der Stufe der Voraktivierung gebildete Komplex $RhH(CO)L_2$, der eigentliche Katalysator ist [45].

Tabelle 11. *Reaktionsgeschwindigkeit r für*

$$S + H_2 \xrightarrow[r]{KaL_2} SH_2$$

mit $KaL_2 = IrX(CO)L_2$ als Funktion von X bei $t = 80\,°C$ in Toluol; $[KaL_2]_0 = 2 \cdot mMol \cdot l^{-1}$

Substrat	$[S]_0$ Mol·l⁻¹	$P(C_6H_{11})_3$ X = Cl	Br	J	$P\varnothing_3$ Cl	Br	J	$P(O\varnothing)_3$ Cl	Br	J
MDME	0,4	0,17	0,044	0,03	$0,27_5$	0,068	0,04	0,36	$0,10_5$	~0,007
MDME	0,8	1,0	0	0	0,47	0,09	0,037	0,53	0,12	~0,01
MA	0,8	19,6	9,8	8,9	3,0	1,37	1,08	0,06	0,03	0
ACAE	0,8	11,2	0	0	10	4,45	2,08	9,3	0,18	0,01
1-Hepten	0,8	$6,2_5$	$5,3_5$	5,3	8,9	3,3	1,6	9,8	1,8	0,18
$\varnothing C \equiv CH$	0,8	0,075	0	0	0,018	0	0	0,36	$0,10_5$	~0,01

Header span: r(mMol·l⁻¹min⁻¹) / für IrX(CO)L₂

h) r als Funktion der Liganden L

Den Einfluß der Liganden L in $IrX(CO)L_2$ auf die Reaktionsgeschwindigkeit r der Hydrierung der Substrate S zeigt Tabelle 12 [10].

Tabelle 12. *Reaktionsgeschwindigkeit r für*

$$S + H_2 \xrightarrow[r]{KaL_2} SH_2 \; mit \; IrX(CO)L_2$$

als Funktion von L bei t = 80 °C in Toluol; $[KaL_2]_0 = 2 \; mMol \cdot l^{-1}$

	$IrX(CO)L_2$		$r(mMol \cdot l^{-1} \cdot min^{-1})$ für Substrate S (c in $Mol \cdot l^{-1}$)					
X	L	MDME $c = 0,4$	MDME $c = 0,8$	MA $c = 0,8$	ACAE $c = 0,8$	1-Hepten $c = 0,8$	$\emptyset C \equiv CH$ $c = 0,8$
Cl	$P(C_6H_{11})_3$	0,17	1,0	19,6	11,2	6,25	$0,07_5$
	$P(iC_3H_7)_3$	0,07	0,06	1,05			
	$P(C_6H_5)_2C_4H_9$	0,15					
	$P(CH_2\emptyset)_3$	0,15					
	$P(p\text{-tolyl})_3$	0,26					
	$P(o\text{-tolyl})_3$	0					
	$P\emptyset_3$	$0,27_5$	0,47	3,0	10,0	8,9	0,018
	$P(O\emptyset)_3$	0,36	0,53	0,06	9,3	9,8	0,36
Br	$P(C_6H_{11})_3$	0,044	0	9,8	6	$5,3_5$	0
	$P(iC_3H_7)_3$	$0,022_5$					
	$P\emptyset_3$	0,068	0,09	1,37	4,45	3,3	0
	$P(O\emptyset)_3$	0,105	0,12	0,03	0,18	1,8	$0,10_5$
J	$P(C_6H_{11})_3$	0,03	0	8,9	0	5,3	0
	$P(i\text{-}C_3H_7)_3$	0					
	$P\emptyset_3$	0,04	0,037	1,08	2,08	1,6	0
	$P(O\emptyset)_3$	0,007	0,01	0	$\sim0,01$	0,18	$\sim0,01$

(Linke Randspalte, vertikal: Zunahme der π-Acceptorstärke)

Offenbar besteht kein eindeutiger Zusammenhang zwischen r und der π-Acceptorstärke der Liganden L. Dieses Ergebnis steht im Gegensatz zu dem Befund, daß r unabhängig vom Substrat S eine eindeutige Funktion der Liganden X ist, und daß weiterhin die Reaktion von $IrX(CO)L_2$ mit H_2 und dem Substrat S in erkennbarer Weise von der π-Acceptorstärke der Liganden L abhängt. Hält man zunächst an der Modellvorstellung fest, daß die Liganden X und L die Elektronendichte am Reaktionszentrum auf einen für die katalytische Aktivität optimalen Wert einjustieren müssen, so kann für die geladenen Liganden X diese Einjustierung praktisch nur über die σ-Bindung in M–X erfolgen, während die ungeladenen Liganden L diesen Effekt über ihre σ-Bindung und die gegenläufige π-Rückgabebindung in $M \underset{\sigma}{\overset{\pi}{\rightleftharpoons}} L$ verursachen. Nun ist aber für die Liganden L nur ihre π-Acceptorstärke bekannt, während es bis heute kein exaktes, experimentell überprüfbares Maß für die Änderung des Beitrages der σ-Bindung in $M \rightleftharpoons L$ bei Variation von L gibt. Sicher ist nur, daß die Lage der ν_{co}-Frequenz in monosubstituierten Metallcarbonylen aus der die π-Acceptorstärke abgeleitet wurde, nicht auf die Stabilität der σ-Bindung in $M \rightleftharpoons L$ und σ-Donatorstärke des Liganden L anspricht, wie die Untersuchungen für L = Stickstoff- oder Sauerstoff-Basen zeigten [46]. Da bei der Komplexbildung des Katalysators mit dem Substrat d_π–p_π-Bindungen beansprucht werden, ist verständlich, daß die Katalysator-Substrat-Gleichgewichte von der π-Acceptorstärke der Liganden L und des Substrates eindeutig abhängen (Tabelle 5). Der Befund, daß die Reaktionsgeschwindigkeit r der Hydrierung nicht eindeutig von der π-Acceptorstärke der Liganden L abhängt, wohl aber von der Änderung der σ-Bindung Ir-X in $IrX(CO)L_2$, zeigt, daß für die Reaktivität des Katalysators die Elektronenverteilung bzw. Elektronendichte in den σ-Bindungen ausschlaggebend ist. Es muß also im Reaktionsmechanismus ein Schritt vorhanden sein, der spezifisch auf die Elektronendichte in den σ-Bindungen anspricht (s. Abschnitt m).

i) r als Funktion des Zentralatoms

Zu diesem Gesichtspunkt liegen nur wenige Untersuchungen vor, welche bei gleicher Katalysatorkonzentration und gleichem Substrat im selben Lösungsmittel durchgeführt wurden. Trotzdem lassen sich aus der Tabelle 13 gewisse Tendenzen ablesen.

Tabelle 13. *Reaktionsgeschwindigkeit r der katalytischen Hydrierung der Substrate S als Funktion des Zentralatoms des Katalysators;* $[KaL_x]_0 = 2 mMol \cdot l^{-1}$; $[S]_0 = 0,8 Mol \cdot l^{-1}$; *Lösungsmittel: Toluol*

t	Katalysator	$r \cdot (mMol \cdot l^{-1} \cdot min^{-1})$; für Substrat $S \equiv$					Lit.
°C		MDME	1-Hepten	Styrol	1-Hexin	ω-Brom Styrol	
80	$IrCl(CO)(P\phi_3)_2$	0,44	8,9	6,66	0	0	[31]
80	$RhCl(CO)(P\phi_3)_2$	0,59	3,04	1,1	0,32	0,14	[31]
25	$IrH(CO)(P\phi_3)_3$	0,225	0,266	0,35	0,133	0,008$_3$	[15]
25	$RhH(CO)(P\phi_3)_3$	0,985	2,85	2,23	1,0	0,025	[15]
25	$RuH(Cl)(P\phi_3)_3$ *		34				[12]
25	$IrCl(P\phi_3)_3$						[47]
25	$RhCl(P\phi_3)_3$	5,3	5,7	7,8	2,45	0,14	[5]

* $[KaL_3]_0 = 8,3 \cdot 10^{-4} Mol \cdot l^{-1}$; $[S]_0 = 1,2 Mol \cdot l^{-1}$ in Benzol

Der Komplex $RhCl(CO)(P\varnothing_3)_2$ ist kein eigentlicher Hydrierungskatalysator, da er erst voraktiviert werden muß, wobei sich $RhH(CO)L_2$ bildet [45], während der analoge Ir-Komplex $IrCl(CO)(P\varnothing_3)_2$ im Temperaturbereich von 70–90 °C ein guter Katalysator ist. Wesentlich aktiver ist der Komplex $IrH(CO)(P\varnothing_3)_3$ mit dem bereits bei 25 °C Hydrierungen möglich sind. Der analoge Rh-Komplex $RhH(CO)(P\varnothing_3)_3$ weist gegenüber dem Ir-Komplex eine ~10 mal höhere Aktivität auf. Das Gleiche gilt für die Katalysatoren vom Typ $MCl(P\varnothing_3)_3$. Auch hier hat der Rh-Komplex eine wesentlich größere Aktivität als der Ir-Komplex. Soweit die spärlichen Untersuchungen einen Schluß zulassen, scheint der aktivste Katalysator der Komplex $RuH(Cl)(P\varnothing_3)_3$ zu sein.

Um einen Ansatzpunkt für eine Erklärung zu finden, wie bei gleichem Typ des Komplexes das Zentralatom die Aktivität ändern kann, sind in der Tab. 14 rein phänomenologisch einige typische Eigenschaften vergleichbarer Komplexe gegenübergestellt. Es zeigt sich, daß die kovalenten Radien der Zentralatome fast identisch sind. Daraus folgt eine höhere Gesamt-Elektronen-Dichte am Iridium. Dies erklärt die geringere Dissoziation der am Iridium gebundenen Liganden und die bessere Komplexbildung mit dem Substrat S im Vergleich zu den entsprechenden Rhodiumkomplexen. Da andererseits die Rhodiumkomplexe der Tabelle 14 wesentlich aktivere Katalysatoren als die Iridiumkomplexe sind, folgt daraus, in Übereinstimmung mit der Modellvorstellung, daß ein bestimmter Katalysator umso aktiver ist, *je schwächer seine Liganden L gebunden sind, und je kleiner die Stabilität des Adduktes aus Katalysator und Substrat ist*, ohne jedoch auf den Wert Null abzusinken.

Tabelle 14. *Moleküleigenschaften der Katalysatoren MCl(P$\varnothing$)$_3$ und MHCO(P$\varnothing_3$)$_3$ und Reaktionsgeschwindigkeit r für die Hydrierung (M = Ir und Rh)*

Moleküleigenschaften der Komplexe	Ir-Komplex	Rh-Komplex
kovalenter Radius M^0	1,36 Å	1,34 Å
Gesamtzahl der Elektronen in M^0	77	45
Elektronen im f-Zustand	14	0
Valenzelektronen von M^0 im Grundzustand	$(5d)^7(6s)^2$	$(4d)^8(5s)^1$
Abstand M-P in $M(H)(CO)(P\varnothing_3)_3$	2,32 Å	2,34 Å
α_{Diss} in $MH(CO)(P\varnothing_3)_3$ nach $KaL_x \rightleftharpoons KaL_{x-1} + L$	<0,02	~1,0
α_{Diss} in $MCl(P\varnothing_3)_3$ nach $KaL_x \rightleftharpoons KaL_{x-1} + L$	~0,23	~0,95
Komplexe aus Ka und Substrat S	$(KaS)_{Ir}$ stabiler	$(KaS)_{Rh}$
r für Hydr. mit Ka $\equiv MCl(P\varnothing_3)_3$	r_{Ir} <	r_{Rh}
mit Ka $\equiv MH(CO)(P\varnothing_3)_3$	r_{Ir} <	r_{Rh}

j) Selektivität der Katalysatoren

Für die Praxis der homogenen katalytischen Hydrierung spielt auch die Frage nach der Selektivität des Katalysators eine Rolle. In der Tabelle 15 sind die Ergebnisse von Untersuchungen, die unter vergleichbaren Bedingungen ausgeführt wurden, angegeben. Als Maß für die Selektivität wurde wieder die Reaktionsgeschwindigkeit r genommen. Für die drei Typen der Katalysatoren $MClL_3$, $MCl(CO)L_2$ und $MH(CO)L_3$ ergeben sich die folgenden Selektivitäten bezüglich der Hydrierung ungesättigter Verbindungen:

a) r ist für endständige Olefine größer als für 2 oder 3-Olefine (Nr. 1-4),

b) r ist für Olefin größer als für Acetylene (Nr. 5-8),

c) Substitution des H durch Br erniedrigt r sehr stark (Nr. 9, 10),

d) r ist für ACAE > MDME > FDME (Nr. 11-13).

Diese Ergebnisse führen zu dem Schluß, daß die *Struktur des Substrates* einen eindeutigen Einfluß auf den Mechanismus der Wasserstoff-Übertragung hat und daß dieser Einfluß in erster Näherung vom Typ des Katalysators unabhängig ist. Betrachtet man die absoluten Werte der Reaktionsgeschwindigkeit r (Tabelle 17, Nr. 16) und die Reaktionstemperatur (Nr. 15) für die Hydrierung von 1-Hepten[a]),

Tabelle 15. *r als Maß für die Selektivität von Hydrierungskatalysatoren* $[Ka]_0 = 2\ mMol \cdot l^{-1}$; $[S]_0 = 0{,}8\ Mol \cdot l^{-1}$; *Lösungsmittel: Toluol*

Nr.	Substrat	$r_{relativ}$ wenn $r_{1\text{-Hepten}}$ = 100 für				
		$IrCl(CO)(P\varphi_3)_2$	$IrH(CO)(P\varphi_3)_3$	$IrCl(P\varphi_3)_3$	$RhH(CO)(P\varphi_3)_3$	$RhCl(P\varphi_3)_3$
1	1-Hepten	100	100	100	100	100
2	2-*cis*-Hepten	11	50	4,8	6,4	27,5
3	2-*trans*-Hepten	6,1	50	2,3	4,9	0,9
4	3-*trans*-Hepten	8,0	28,3	5,4	4,7	0,5
5	1-Hepten	100	100	100	100	100
6	1-Hexin		50	4,3	35	43,5
7	Styrol	73,4	132	24	78,3	138
8	Phenylacetylen	2,0	56,5	3,5	39,2	18
9	Styrol	73,4	132	24	78,3	138
10	ω-Bromstyrol	0	3,1	6	0,9	2,5
11	ACAE	111	100	121	260	152
12	MDME	5	84,8	42	34	94
13	FDME	0	18,9	0	3,7	70
14	Lit.	31)	15)	47)	15)	5)
15	Temp. °C	80	25	25	25	25
16	r (mMol·l^{-1}·min^{-1}) für 1-Hepten	8,93	0,265	0,65	0,85	5,66

[a]) r für 1-Hepten mit $IrCl(CO)(P\varphi_3)_2$ ist bei 25 °C unmeßbar klein.

so ergibt sich aus der Tabelle 15 und unter Einbeziehung der Ergebnisse mit
$RuH(Cl)(P\emptyset_3)_3$ [12)] für die Reaktivität der Katalysatoren die Reihe:

$IrCl(CO)(P\emptyset_3)_2 < IrH(CO)(P\emptyset_3)_3 < IrCl(P\emptyset_3)_3$

$< RhH(CO)(P\emptyset_3) < RhCl(P\emptyset_3)_3 < RuH(Cl)(P\emptyset_3)_3$

Bezüglich der *Selektivität* ergibt Tabelle 15, daß sie für die verschiedenen Olefine
(Nr. 1–4) und von Olefinen verglichen mit Acetylenen (Nr. 5–8) umso ausgeprägter ist, je größer der Absolutwert von r für die Hydrierung von 1-Hepten
(Nr. 16) unter den gegebenen Reaktionsbedingungen ist. Die größte Selektivität besitzt $RuH(Cl)(P\emptyset_3)_3$, da sich hier r für 1-Hexen und 2-Hexen wie 1000:1
verhält [12)]. Ein detaillierteres Problem wäre die Frage nach der Änderung der
Selektivität bei gleichem Katalysatortyp und Zentralatom als Funktion der Liganden X und L. Hierzu liegen nur wenig systematische und vergleichbare Untersuchungen vor. Schreibt man jedoch Tabelle 11 in die Form der Tabelle 16 um
und setzt wieder für $r_{1\text{-Hepten}} = 100$, so erkennt man, daß in den Komplexen
$IrX(CO)L_2$ bei beibehaltenem Liganden L jeweils die Selektivität des Katalysators gegenüber dem Substrat S eindeutig ist. Dies ist letzlich eine Folge der ex-

Tabelle 16. *Selektivität von IrX(CO)L$_2$ als Funk-
tion von X und L. $[Ka]_0 = 2 mMol \cdot 1^{-1}$;
$[S]_0 = 0,8 Mol \cdot 1^{-1}$; $t = 80\,°C$; Lösungsmittel:
Toluol ($r_{1\text{-Hepten}} = 100$) [48)]*

	X =	Cl	Br	J
Nr.	Substrat S	L = P(C$_6$H$_{11}$)$_3$		
1	1-Hepten	100	100	100
2	MA	315	183	168
3	ACAE	179	0	0
4	MDME	16	0	0
5	$\emptyset C\equiv CH$	1,2	0	0
		L = P$\emptyset_3$		
1	1-Hepten	100	100	100
3	ACAE	112	135	130
2	MA	33,5	41,5	67,5
4	MDME	5,2	2,7	2,3
5	$\emptyset C\equiv CH$	0,2	0	0
		L = P(O$\emptyset$)$_3$		
1	1-Hepten	100	100	100
3	ACAE	95	10	5,6
4	MDME	5,4	6,7	5,6
5	$\emptyset C\equiv CH$	3,7	5,8	5,6
2	MA	0,06	1,7	0

perimentellen Erfahrung, daß r für X = Cl > Br > J ist. Durch Variation der Liganden X läßt sich also der absolute Wert von r variieren, die relative Abstufung der r-Werte für die Substrate S der Tabelle 16 bleibt jedoch erhalten. Variiert man jedoch den Liganden L im Katalysator $IrX(CO)L_2$ [48], so tritt eine gravierende Änderung in der Selektivität gegenüber den Substraten S auf, welche wieder unabhängig vom Liganden X ist (Tabelle 16). Es gilt daß mit:

L = $P(C_6H_{11})_3$ r: für 1-Hepten < MA > ACAE > MDME > $\emptyset C \equiv CH$

L = $P\emptyset_3$ r: für 1-Hepten < ACAE > MA > MDME > $\emptyset C \equiv CH$

L = $P(O\emptyset)_3$ r für 1-Hepten > ACAE > MDME > $\emptyset C \equiv CH$ > MA ist.

Diese vorläufigen Ergebnisse, welche unbedingt weiter ausgebaut werden müßten, zeigen, daß nur durch Variation der Liganden L des Katalysators KaL_x die Selektivität gegenüber vorgegebenen Substraten S prinzipiell geändert werden kann. Für die Praxis würde dies z.B. bedeuten, daß in einem Gemisch von MA und MDME mit $IrCl(CO)[P(O\emptyset)_3]_2$ praktisch nur MDME hydriert wird, während mit $IrCl(CO)[P(C_6H_{11})_3]_2$ vorzugsweise MA hydriert wird (s.Tabelle 16). Sind solche Selektivitätsunterschiede für verschiedene Substrate S als Funktion von L gefunden, so kann, wie Tabelle 16 zeigt, durch Variation des geladenen Liganden der Effekt wesentlich verstärkt werden. So hydriert der Jodo-Komplex $IrJ(CO)[P(O\emptyset)_3]_2$ nur MDME und $IrJ(CO)[P(C_6H_{11})_3]_2$ nur MA aus einem Gemisch von MA und MDME.

k) r und Katalysatorkonzentration $[KaL_x]_0$

Vom theoretischen Aspekt der Kinetik erhält man aus der Variation von r mit der Einwaagekonzentration des Katalysators $[KaL_x]_0$ die Information, ob der eingesetzte Komplex der eigentliche Katalysator ist oder ein mit ihm im Gleichgewicht stehendes Folgeprodukt. Für die Praxis der katalytischen Reaktionsführung ist die entscheidende Frage, wie man mit einem Minimum an Katalysator eine optimale Umsatzgeschwindigkeit erhält? Aus den Gleichungen II bis V der Tabelle 3 für die kinetischen Ansätze der katalytischen Hydrierung folgt, daß nach Gleichung VII

$$r \text{ prop } [KaL_x]_0 \qquad\qquad VII$$

die Reaktionsgeschwindigkeit r proportional der Einwaagekonzentration $[KaL_x]_0$ sein müßte, wenn die Substratkonzentration konstant gehalten wird. Erste quantitative Untersuchungen mit $RhCl(P\emptyset_3)_3$ [22] und $RhH(CO)(P\emptyset_3)_3$ [24] in Benzol zeigten, daß diese Beziehung nur im Konzentrationsbereich $[KaL_x]_0$ = 0,5 bis 2,0 bzw. 1,5 bis 5,0 mMol $\cdot$ l^{-1} gilt. Spätere Untersuchungen in Toluol, die bis zu $[KaL_x]_0$ = 8 mMol $\cdot$ l^{-1} ausgedehnt wurden, ergaben im vermessenen Konzentrationsbereich keine Linearität von $r = f[KaL_x]_0$ [25,26]. Es konnte nun für den Katalysator $RhH(CO)(P\emptyset_3)_2$ gezeigt werden, daß im gesamten Konzentrationsbereich das Gleichgewicht des Reaktionsschemas

W. Strohmeier

$$RhH(CO)(P\emptyset_3)_2 \rightleftharpoons RhH(CO)(P\emptyset_3) + P\emptyset_3 \qquad (16)$$

vorliegt [25]. Fügt man dem Hydrierungssystem jedoch freien Liganden L zu, so werden von einer Grenzkonzentration ab die r-Werte prop $[KaL_x]_0$, da dann das Gleichgewicht (16) völlig links liegt und man in der Lösung nur den weniger aktiven Katalysator $RhH(CO)(P\emptyset_3)_2$ vorliegen hat, dessen Aktivität man dann bestimmt [25]. Abb. 2 zeigt die Abnahme von r bei Zugabe von freiem Liganden L und die gemessene Proportionalität von r mit $[KaL_x]_0$. Analog liegen die Verhältnisse bei den Komplexen $RhX(P\emptyset_3)_3$, die, wie die Tabelle 10 zeigt, ebenfalls zwei Liganden abdissoziieren können und dabei die aktiveren Spezies $RhX(P\emptyset_3)$ bilden [51]. In der Praxis wird man natürlich ohne Ligandenzusatz bei möglichst kleiner $[KaL_x]_0$ arbeiten, da mit abnehmender Katalysatorkonzentration der Quotient $r/[KaL_x]_0$ immer größer wird. Weiterhin sollte auch r in Lösungsmitteln, welche die Dissoziation begünstigen, größer werden, was bestätigt wurde.

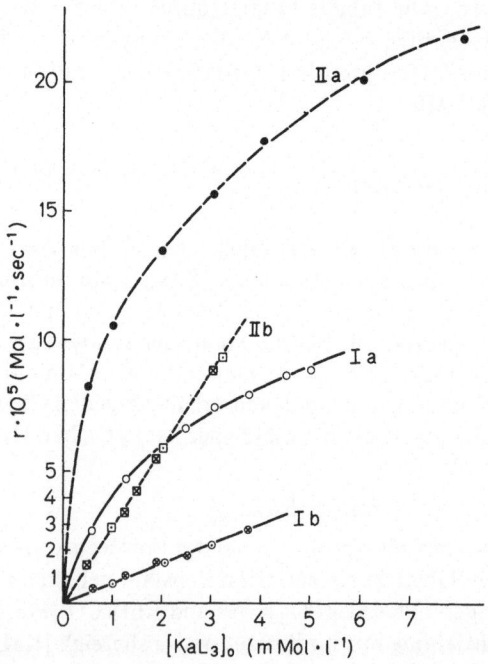

Abb. 2. Reaktionsgeschwindigkeit r für die Hydrierung von MDME und 1-Hepten als Funktion von $[RhHCO(P\emptyset_3)_3]_0$ ohne und mit Zusatz von L = $P\emptyset_3$. Lösungsmittel: Toluol: $[S]_0 = 0,6 \cdot Mol \cdot l^{-1}$
I: 1-Hepten bei $t = 25\,°C$. a) $[P\emptyset_3]_0 = 0$. b) $[P\emptyset_3]_0 = 3,75$ ⊕ und 6 mMol $\cdot l^{-1}$ ⊙
II: MDME bei $t = 70\,°C$. a) $[P\emptyset_3]_0 = 0$. b) $[P\emptyset_3]_0 = 60$ ◻ und 90 mMol $\cdot l^{-1}$ ◼

So ist r für die Hydrierung von Cyclohexen mit $RhCl(P\emptyset_3)_3$ in C_2H_5OH > $CH_3COC_2H_5$ > Benzol [22] und in Benzol > (Benzol/Heptan) [23].

I) r als Funktion der Substratkonzentration

Quantitative Messungen zur Abhängigkeit der Reaktionsgeschwindigkeit r der katalytischen Hydrierung von der Substratkonzentration $[S]_0$ in Benzol mit $RhCl(P\emptyset_3)_3$ [22] und $RhH(CO)(P\emptyset_3)_3$ [24] zeigten, daß r mit steigender Substratkonzentration asymptotisch einem Grenzwert zustrebt.

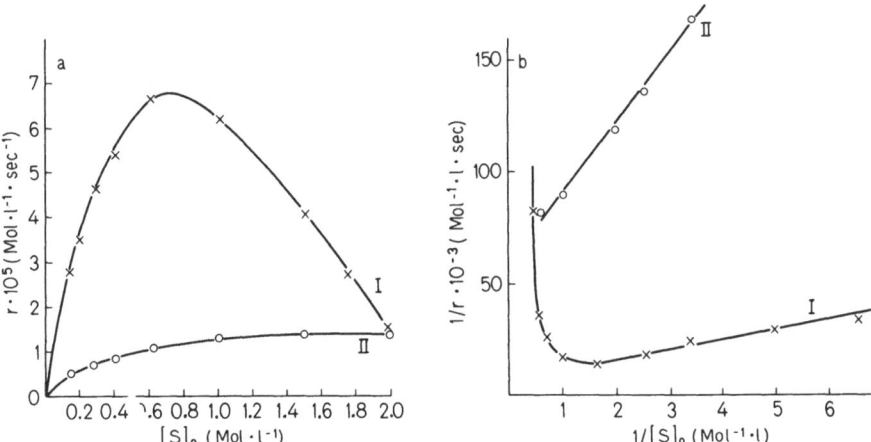

Abb. 3a u. b. Funktionelle Zusammenhänge zwischen der Reaktionsgeschwindigkeit r der Hydrierung und der Substratkonzentration $[S]_0$ [24,25]. a) r gegen $[S]_0$. b) $1/r$ gegen $1/[S]_0$. $[RhH(CO)(P\emptyset_3)_3]_0 = 1,25$ mMol $\cdot 1^{-1}$; $t = 30\,°C$
I: 1-Hepten in Toluol   II: 1-Hepten in Toluol plus $P\emptyset_3$ (3,75 mMol $\cdot 1^{-1}$)

Durch Umformen der Gleichungen II bis V der Tab. 3 ergibt sich, daß

$$1/r \text{ prop } 1/[S]_0 \qquad\qquad VII$$

ist, wenn $[KaL_x]_0$ konstant gehalten wird. In Benzol wurde nun diese Linearität von $1/r$ mit $1/[S]_0$ im Konzentrationsbereich von $[S]_0 = 0,5$ bis $2,5$ Mol $\cdot 1^{-1}$ auch nachgewiesen [23,24]. Messungen in Toluol ergeben jedoch (Abb. 3a, Kurve I), daß mit steigender Substratkonzentration die Reaktionsgeschwindigkeit r zunächst stark zunimmt, dann durch eine Maximum geht und bei hohen Werten von $[S]_0$ abfällt [25], und $1/r$ gegen $1/[S]_0$ keine Gerade gibt (Abb. 3b, Kurve I). Durch Zugabe von freiem Liganden $P\emptyset_3$ wird zwar r stark erniedrigt, da der aktive Katalysator $RhH(CO)P\emptyset_3$ in $RhH(CO)(P\emptyset_3)_2$ überführt wird, aber r nähert sich jetzt mit zunehmender Substratkonzentration einem Grenzwert.

(Abb. 3 a, Kurve II), und bei nicht extrem hohen Substratkonzentrationen wird $1/r$ prop $1/[S]_0$ (Abb. 3 b, Kurve II). Der Effekt, daß r mit zunehmender Substratkonzentration ohne Zusatz von $P\emptyset_3$ durch ein Maximum geht, scheint in Toluol viel ausgeprägter als in Benzol zu sein und läßt sich zwanglos mit der Annahme erklären, daß nach Schema (17)

$$RhH(CO)(P\emptyset_3) \; \overset{S}{\rightleftharpoons} \; RhH(CO)(P\emptyset_3)S \; \overset{S}{\rightleftharpoons} \; RhH(CO)(P\emptyset)S_2 \qquad (17)$$

der intermediäre Komplex $RhH(CO)P\emptyset_3S$ bei sehr hohen Substratkonzentrationen den Komplex $RhH(CO)P\emptyset_3S_2$ bildet, der keinen Wasserstoff mehr aufnehmen und auf das Substrat übertragen kann [25].

m) Zur Frage der H_2-Übertragung

Das interessanteste Problem bei den homogenen katalytischen Hydrierungen ist die Beantwortung der Frage, nach welchem *Mechanismus* die Wasserstoffübertragung geschieht. Unabhängig davon ob die Katalyse über die *hydride* oder *unsaturate route* verläuft, muß als intermediäres Produkt ein Komplex KaH_2S gebildet werden, der sowohl H_2 als auch das Substrat S angelagert hat. Der Mechanismus läßt sich am einfachsten mit dem Katalysator $RhClL_3$ darstellen [22,52].

Es bildet sich intermediär die Rhodium-Alkylverbindung a. Versuche zur Isomerisierung mit diesem Katalysator in Gegenwart von Deuterium [52] führen zu dem Schluß, daß die Wasserstoff-Übertragung in einem Zwei-Schritt-Mechanismus abläuft. Der direkte Nachweis einer Rhodiumalkyl-Verbindung ist noch nicht geglückt und wird schwer zu bringen sein, da der zweite Schritt der H-Übertragung im Komplex energetisch sicher günstig liegt. Bessere Chancen für den Nachweis intermediärer Alkylverbindungen sollten bei Hydrierungskatalysatoren wie $IrH(CO)L_3$ ($\equiv IrHL_x$) vorhanden sein, da hier der erste Schritt der H-Übertragung unter Benutzung des bereits im Komplex vorhandenen Hydridwasserstoffes nach Schema (18) möglich ist [16].

$$L_x IrH + C_2H_4 \overset{H}{\rightleftharpoons} L_x IrC_2H_4 \rightleftharpoons L_x IrC_2H_5 \overset{H_2}{\longrightarrow} L_x IrC_2H_5 \overset{H_2}{\longrightarrow} L_x IrH + C_2H_6 \qquad (18)$$

Für die Platinkomplexe $PtHX[P(C_2H_5)_3]_2$ konnte durch Isolierung der Alkylverbindung $Pt(C_2H_5)X[P(C_2H_5)_3]_2$ bewiesen werden, daß Äthylen nach Schema (19) mit der Alkylverbindung im Gleichgewicht steht [53].

$$PtHX[P(C_2H_5)_3]_2 \; + \; C_2H_4 \; \rightleftharpoons \; Pt(C_2H_5)X[P(C_2H_5)_3]_2 \qquad (19)$$

Tabelle 17. Substrate (S), welche auf eine katalytische Hydrierung mit Komplexverbindungen der 8. Gruppe überprüft wurden (die Zahlen geben die Literaturstellen an)

Nr.	Typ	Substrat S	$RhXL_3$	$RhX(CO)L_2$	$RhH(CO)L_3$	$IrX(CO)L_2$	$IrH(CO)L_3$	$RuHClL_3$
1	1-Olefine	1-Penten	+ 4,23,29)		+ 24)			+ 61)
2		1-Hexen	+ 27,29)	+ 8,31,60)	+ 15,25)			+ 61)
3		1-Hepten	+ 27)			+ 31)	+ 15)	+ 61)
4		1-Octen	+ 27)					
5		1-Nonen	+ 27)					
6		1-Decen	+ 27)		+ 24)			+ 61)
7		1-Undecen	+ 27)		+ 24)			
8		1-Dodecen	+ 27)					
9		2-Methyl-1-penten	+ 23,27)					− 61)
10		2,4,4-trimethyl-1-penten	+ 27)					
11		Styrol	+ 23,29)	+ 8,31,60)	+ 15,24)	+ 31)	+ 15)	
12		4-Methoxystyrol	+ 23)					
13		4-Fluorstyrol	+ 23)					
14		2-Methyl-1-penten			− 24)			+ 61)
15	2-Olefine	2-cis-penten	+ 23)		− 24)			
16		2-Hexen			− 24)			− 61)
17		2-cis-Hepten	+ 29)	+ 8,31,60)	+ 15) −24)	+ 31)	+ 15)	
18		2-trans-Hepten	+ 29)	+ 8,31,60)	+ 15)	+ 31)	+ 15)	
19		2-Octen	+ 27)					− 61)
20		3-Äthyl-2-penten	+ 23)					
21		2,3-Dimethyl-2-buten	+ 23)					
22		4-Methyl-cis-2-penten	+ 4,23)		− 24)			
23		4-Methyl-trans-2-penten	+ 4,23)					
24	3-Olefine	3-trans-Hexen	+ 23)					
25		3 cis-trans-Hepten		+ 8)				
26		3-trans-Hepten	+ 29)	+ 8,31)	+ 15)	+ 31)	+ 15)	− 61)

Tabelle 17 (Fortsetzung)

Nr.	Typ	Substrat S	RhXL₃	RhX(CO)L₂	RhH(CO)L₃	IrX(CO)L₂	IrH(CO)L₃	RuHClL₃
27		Cyclopenten	+ 23,27)					
28		Cyclohexen	+ 4,23,27)		− 24)			− 61)
29		Cyclohepten	+ 27,29)	+ 8,31)	+ 15)	+ 31)	+ 15)	
30		Cyclooocten	+ 27)	+ 60)				
31		Norbonen		+ 60)				
32		1-Methylcyclohexen	+ 23,59)					
33		1,4-Dimethylcyclohexen	+ 59)					
34	Cycloolefine	4-Vinylcyclohexen			+ 24)			
35		Isopren	+ 27)					
36		1-3-Pentadien	+ 23)		− 24)			+ 12)
37		1-5-Hexadien	+ 23,27)		+ 24)			
38		1-3-Octadien	+ 27)					
39		1-4 trans-Hexadien						+ 61)
40		1-4 cis-Hexadien						+ 61)
41	Diene	2-Methyl-1-5-hexadien						+ 61)
42		1,4-Cyclohexadien	− 23)	+ 60)				
43	Cyclo-diene	1,3-Cyclohexadien	+ 23)		+ 24)			
44		1,3-Cyclooctadien	+ 27)	− 60) + 8)	+ 15)		+ 15)	
45		Allylcyanid	+ 23)		+ 24)			
46		Allylalkohol	+ 23)		+ 24)			
47		Allylchlorid	− 23)					
48		Allylacetat	+ 27)					
49		Allylbenzol			+ 24)			
50	Allylverbindungen	Zimtaldehyd			− 24)			

Tabelle 17 (Fortsetzung)

Nr.	Typ	Substrat S	$RhXL_3$	$RhX(CO)L_2$	$RhH(CO)L_3$	$IrX(CO)L_2$	$IrH(CO)L_3$	$RuHClL_3$
51	ungesättigte Äther	Allyl-phenyläther	− 23)		− 24)			
52		iso-Butylvinyläther	+ 27)					
53		n-Butylvinyläther		+ 60)				
54	ungesättigte Säuren und Derivate	Acrylsäure	− 23)		− 24)			
55		Acrylsäureäthylester	+ 29)	+ 8,31)	+ 15)	+ 31)	+ 15)	
56		Methylmetacrylat	+ 27)					
57		Acrylnitril	+ 27)					
58		Acrylamid	+ 23)					
59		Maleinsäureanhydrid	− 23)		+ 15)			
60		Maleinsäuredimethylester	+ 29)	+ 8,31,60)	+ 15)	+ 31)	+ 15)	
61		Fumarsäuredimethylester	+ 29)	+ 8,31)	+ 15)	− 31)	+ 15)	
62	Halogen-Olefine	1-Chlor-1-propen			− 24)			
63		2-Chlor-1-propen			− 24)			
64		Tetrafluoräthylen	− 23)					
65		Tetrachloräthylen	− 23)					
66		ω-Bromstyrol	+ 29)	+ 8,31)	+ 15)	− 31)	+ 15)	
67	Acetylene und Derivate	1-Hexin	+ 22,27,29)	+ 8,31)	+ 15-24)	− 31)	+ 15)	
68		1-Heptin	+ 27)					
69)		1-Octin	+ 27)					
70		Phenylacetylen	+ 27,29)	+ 8,31)	+ 15)	+ 31)	+ 15)	
71		Diphenylacetylen	+ 62)					
72		3-methyl-1-pentin-3-ol	+ 62)					
73		3-methyl-butin-3-ol	+ 62)					
74		2-pentin-1-4-diol	+ 62)					
75		1-Äthinylcyclohexan-1-ol	+ 62)					
76		3-Chlorpropin	− 62)					
77		Acetylendicarbonsäure	− 62)					

Mit Propylen als Olefin liegt das Gleichgewicht (19) stark links und mit 1-Octen konnte keine Alkylverbindung mehr nachgewiesen werden [53].

Die Untersuchungen des Gleichgewichts (19) als Funktion der Liganden L (L ≡ PR$_3$) ergäben einen Hinweis, wie die Bildung der Alkylverbindung von den Eigenschaften der Liganden abhängt. Es ist durchaus möglich, daß der geschwindigkeitsbestimmende Schritt der katalytischen Hydrierung in der Bildung der σ-Alkyl-Metallbindung liegt. Der experimentelle Befund, daß die Reaktionsgeschwindigkeit r nicht eindeutig von der π-Acceptor-Stärke der Liganden L abhängt (Abschnitt VII. h), könnte damit zusammenhängen, daß für die Bildung der σ-Alkyl-Metallbindung nach (19) die Elektronendichte am Zentralatom über die σ-Metall-Phosphorbindung auf einen optimalen Wert einjustiert werden muß. Wie erwähnt, ist jedoch die Änderung der Elektronendichte in der M-L-σ-Bindung als Funktion der Liganden L nicht bekannt.

n) Isomerisierung während der Hydrierung

Es war bekannt, daß Olefine durch Übergangsmetallkomplexe isomerisiert werden können [54,58]. Es zeigt sich nun, daß auch Hydrierungskatalysatoren wie IrCl(CO)(PØ$_3$)$_2$ in Gegenwart von H$_2$ Isomerisierungskatalysatoren sind [21]. Somit muß bei der katalytischen Hydrierung mit Komplexverbindungen gleichzeitig mit einer Isomerisierung gerechnet werden, was, wie diesbezügliche Untersuchungen ergaben, auch der Fall ist [55,56]. Dieser Effekt muß unbedingt bei der quantitativen Auswertung kinetischer Daten zur Hydrierung berücksichtigt werden, da die Isomerisierungsgeschwindigkeit in der Größenordnung der Hydrierungsgeschwindigkeit liegen kann und in manchen Systemen größer als die Hydrierungsgeschwindigkeit ist [55]. Weitere Versuche ergaben, daß die Isomerisierungsgeschwindigkeit unter den Bedingungen der Hydrierung unter H$_2$ wesentlich größer ist [55] als z.B. unter Stickstoff [57]. Bei quantitativen Untersuchungen zur Kinetik muß somit immer die Reaktionsgeschwindigkeit r *zu Beginn* der Hydrierung als Maß für die eigentliche Aktivität des Katalysators genommen werden. Für die Praxis der Hydrierung ist die Verwendung von Katalysatoren, welche eine kleine Isomerisierungsgeschwindigkeit haben, vorteilhaft, da, wie die Tabelle 15 zeigt, die 2- und 3-Olefine wesentlich langsamer hydriert werden als die 1-Olefine. Der sehr aktive Katalysator RuClH(PØ$_3$)$_3$ isomerisiert während der Hydrierung 1,4-Hexadien und 2-Octen nur sehr langsam [61]. Er besitzt somit hohe Aktivität und Selektivität bei nur geringer Isomerisierung.

VIII. Qualitative Untersuchungen zur Hydrierung ungesättigter Verbindungen

Um Anhaltspunkte zu finden, welche Komplexe potentielle Hydrierungskatalysatoren bezüglich eines vorgegebenen Substrates sein können, enthält die Tabelle 17 eine Zusammenstellung der bisher untersuchten Systeme. Ein + bedeutet, daß unter den in der Literatur angegebenen Bedingungen eine Hydrierung möglich war, ein –, daß keine Hydrierung beobachtet wurde oder daß $r < 0,04$ (mMol·l^{-1}·min^{-1}) war [24,61]. Wie jedoch eingangs erwähnt, besteht durchaus die Möglichkeit, durch schärfere Reaktionsbedingungen, also höhere Temperatur und höheren H_2-Druck, die Hydrierung doch zu ermöglichen. So wird z.B. 1,3-Cyclooctadien mit RhX(CO)L$_2$ bei 70 °C nicht hydriert [60], wohl aber bei 80 °C [8]. Analog verhält sich Cycloocten [60]. Substrate, welche mit einem Komplex nicht getestet wurden sind durch leere Felder gekennzeichnet. Durch Anlogieschluß gelingt es leicht einen potentiellen Katalysator zu finden. So wird mit großer Wahrscheinlichkeit 1-Octen (Tabelle 17, Nr. 4) in Analogie zu 1-Hepten auch mit den Komplexen RhX(CO)L$_2$, RhH(CO)L$_3$, IrXL$_3$, IrX(CO)L$_3$, IrH(CO)L$_3$ und RuHClL$_3$ hydriert werden können.

IX. Zusammenfassung

Aus den bisherigen Untersuchungen über die Komplexverbindungen der Metalle der 8. Gruppe als homogene Hydrierungskatalysatoren ergeben sich die folgenden Gesichtspunkte und Hinweise:

1) Für eine Reihe von Komplexverbindungen des Typs MX$_n$L$_x$Y$_m$ wurde nachgewiesen, daß sie homogene Hydrierungskatalysatoren für ungesättigte Verbindungen bei milden Bedingungen sind (Tabelle 1).

2) Weitere Komplexverbindungen sind als potentielle Katalysatoren zu betrachten, welche bei schärferen Bedingungen hydrieren.

3) Vom Mechanismus her gesehen verläuft die Hydrierung zum überwiegenden Teil über die *„unsaturate route"*.

4) Der Katalysator muß entweder eine unbesetzte Koordinationsstelle besitzen an welche das Substrat S angelagert wird oder

5) er muß das Substrat unter Erhöhung seiner Koordinationszahl anlagern können.

6) Für die Reaktivität spezieller Komplexverbindungen als Hydrierungskatalysatoren wurde mit 1-Hepten als Substrat die folgende Abstufung gefunden:

$$IrCl(CO)(P\phi_3)_2 < IrH(CO)(P\phi_3)_3 < IrCl(P\phi_3)_3 < RhH(CO)(P\phi_3)_3 < RhCl(P\phi_3)_3 < RuH(Cl)(P\phi_3)$$

7) Bei gleichem Typ der Komplexverbindung hat der Rhodiumkomplex eine wesentlich größere Aktivität als der Iridiumkomplex. Dieser Effekt kann unter anderem damit erklärt werden, daß in den Rhodiumkomplexen die Liganden L nach $KaL_x \rightleftharpoons KaL_{x-1} + L$ leichter abdissoziieren.

8) Die unter 6) genannten Katalysatoren besitzen Selektivität. Die Reaktionsgeschwindigkeit r für die Hydrierung ist für

W. Strohmeier

$r_{1\text{-Olefin}} \gg r_2$- oder $r_{3\text{-Olefin}}$

$r_{\text{Olefin}} > r_{\text{Acetylene}}$

$r_{\text{Olefine}} > r_{\text{Brom-Olefine}}$

9) Durch Variation des geladenen Liganden X in $MX_nL_xY_m$ kann die Reaktionsgeschwindigkeit r gezielt geändert werden, die Selektivität gegenüber vorgegebenen Substraten bleibt jedoch erhalten (Tabelle 16).

10) Durch Variation der ungeladenen Liganden L in $MX_nL_xY_m$ wird r ebenfalls geändert. Gleichzeitig ändert sich die Selektivität gegenüber vorgegebenen Substraten (Tabelle 16).

11) Während r eindeutig von den Liganden X abhängt, wurde kein eindeutiger Zusammenhang zwischen r und der π-Acceptorstärke der Liganden L gefunden.

12) Der Hydridwasserstoff des Komplexes KaL_xH_2S wird im 1. Schritt auf das angelagerte Substrat S unter Bildung einer Metall-Alkyl-Bindung übertragen. Möglicherweise ist diese Reaktion der geschwindigkeitsbestimmende Schritt in der Hydrierung.

13) Für die Übertragung des Hydridwasserstoffes auf das Substrat unter Bildung einer Metall-Alkyl-Bindung muß die Elektronendichte am Reaktionszentrum (Zentralatom) für die σ-Bindung von den Liganden X und/oder L auf einen optimalen Wert einjustiert werden. Für die Liganden X kann dies abgeschätzt werden, für die Liganden L nicht, da der Beitrag der σ-Bindung in $M \rightleftharpoons L$ als Funktion von L nicht bekannt ist.

14) Die Bildung des Adduktes aus Katalysator und H_2 und aus Katalysator und Substrat hängt in eindeutiger Weise von den Liganden X und der π-Acceptorstärke der Liganden L ab.

15) Entgegen den Aussagen der kinetischen Ansätze ist in einem größeren Konzentrationsbereich des Katalysators r nicht linear mit $[KaL_x]_0$, da der Katalysator in verdünnteren Lösungen auch einen 2. Liganden L abdissoziieren kann, wobei ein reaktiverer Komplex entsteht.

16) Für große Konzentrationsbereiche des Substrates S ist auch $1/r$ nicht linear mit $1/[S]_0$, da wahrscheinlich bei sehr hohen Substratkonzentrationen unbesetzte Koordinationsstellen des Komplexes, welche den Wasserstoff aufnehmen müssen, durch das Substrat blockiert werden.

17) Unter den Bedingungen der homogenen katalytischen Hydrierung ungesättigter Verbindungen tritt immer Isomerisierung auf.

X. Literatur

[1] Itatani,H., Bailar, J. C.: J. Am. Chem. Soc. *89*, 1600 (1967).
[2] Bailar, J. C., Itatani, H.: J. Am. Chem. Soc. *89*, 1592 (1967).
[3] Schwab, G.-M., Mandre, G.: J. Catalysis *12*, 103 (1968).
[4] O'Connor, Ch., Wilkinson. G.: Tetrahedron Letters *18*, 1375 (1969).

[5] Strohmeier, W., Endres, R.: Z. Naturforschung *25b*, 1068 (1970).
[6] Bennett, M. A., Milner, D. L.: Chem. Commun. *581* (1967).
[7] Vaska, L., Rhodes, R. E.: J. Am. Chem. Soc. *87*, 4970 (1965).
[8] Strohmeier, W., Rehder-Stirnweiß, W.: Z. Naturforschung *266*, 61 (1971).
[9] James, B. R., Memon, N. A.: Can. J. Chem. *46*, 217 (1968).
[10] Strohmeier, W., Fleischmann, R., Onoda, Takeru: J. Organometal. Chem. *28*, 281 (1971).
[11] Collmann, J. P., Hoffmann, N. W., Morris, D. E.: J. Am. Chem. Soc. *91*, 5969 (1969).
[12] Hallmann, P. S., Evans, D., Osborn, J. A., Wilkinson, G.: Chem. Commun. 305 (1967).
[13] Hidai, M., Kuse, T., Hikita, T., Uchida, Y., Misono, A.: Tetrahedron Letter *20*, 1715 *(1970)*.
[14] O'Connor, Ch., Yagupsky, G., Evans, D., Wilkinson, G.: Chem. Commun. 420 (1968).
[15] Strohmeier, W., Hohmann, S.: Z. Naturforschung *25 b*, 1309 (1970).
[16] Vaska, L.: Inorg. Nucl. Chem. Letters *1*, 89 (1965).
[17] Evans, D., Osborn, J. A., Jardine, F. H., Wilkinson, G.: Nature *208*, 1203 (1965).
[18] James, B. R., Rempel, F. T. T., Rempel, G. L.: Inorg. Nucl. Chem. Letters *4*, 197 (1968).
[19] Coffey, R. S.: Chem. Commun. 923 (1967).
[20] Itatani, H., Bailar, J. C.: J. Am. Oil Chemists' Soc. *44*, 147 (1967).
[21] Eberhardt, G. G., Vaska, L.: J. Catalysis *8*, 183 (1967).
[22] Osborn, J. A., Jardine, F. H., Young, J. F., Wilkinson, G.: J. Chem. Soc. (London) Ser. A. 1711 (1966).
[23] Jardine, F. H., Osborn, J. A., Wilkinson, G.: J. Chem. Soc. (London) Ser. A 1574 (1967).
[24] O'Connor, Ch., Wilkinson, G.: J. Chem. Soc. (London) Ser. A 2665 (1968).
[25] Strohmeier, W., Rehder-Stirnweiß, W.: Z. Naturforschung *26b*, 193 (1971).
[26] – Onoda, Takeru: Z. Naturforschung *24b*, 1493 (1969).
[27] Candlin, J. P., Oldham, A. R.: Discussions Faraday Soc. *46*, 60 (1968).
[28] Strohmeier, W., Endres, R.: Z. Naturforschung *26 b*, 362 (1971).
[29] – – nicht publizierte Messungen.
[30] – Onoda, Takeru: Z. Naturforschung *24b*, 461 (1969).
[31] – Rehder-Stirnweiß, W., Fleischmann, R.: Z. Naturforschung *25 b*, 1481 (1970).
[32] – Structure and Bonding *5*, 96 (1968).
[33] – Müller, F. J.: Chem. Ber. *100*, 2812 (1967).
[34] – Fleischmann, R.: Z. Natruforschung *24 b*, 1217 (1969).
[35] Baddley, W. H.: J. Am. Chem. Soc.: *88*, 4545 (1966).
[36] Strohmeier, W., Onoda, Takeru: Z. Naturforschung *24b*, 515 (1970).
[37] Chock, P. B., Halpern, J.: J. Am. Chem. Soc. *88*, 3511 (1966).
[38] Strohmeier, W., Endres, R.: nicht publizierte Messungen.
[39] Bath, S. S., Vaska, L.: J. Am. Chem. Soc. *85*, 3500 (1963).
[40] Maque, J. T., Wilkinson, G.: J. Chem. Soc. (London) Ser. A 1736 (1966).
[41] Evans, D., Yaguspsky, G., Wilkinson, G.: J. Chem. Soc. (London) Ser. A 2660 (1968).
[42] Strohmeier, W., Rehder-Stirnweiß, W., Reischig, G.: J. Organometal. Chem. *27*, 393 (1971).
[43] Bennett, M. A., Longstaff, P. A.: Chem. & Ind. 846 (1965).
[44] Baird, M. C., Wilkinson, G.: J. Chem. Soc. (London) Ser. A 865 (1967).
[45] Strohmeier, W., Rehder-Stirnweiß, W., Fleischmann, R.: Z. Naturforschung *25b*, 1480 (1970).
[46] – Guttenberger, J. F., Hellmann, H.: Z. Naturforschung *19b*, 353 (1964).
[47] – Endres, R.: Z. Naturforschung *26b*, 730 (1971).
[48] – Fleischmann, R.: J. Organometal. Chem. *29*, C 39 (1971).
[49] Onoda, Takeru: Dissertation Universität Würzburg, 1969.
[50] Strohmeier, W.: J. Organometal. Chem. *32*, 137 (1971).
[51] – Endres, R.: nicht publizierte Messungen.
[52] Biellmann, J. F., Jung, M. J.: J. Am. Chem. Soc. *90*, 1673 (1968).
[53] Chatt, J., Coffey, R. S., Gough, A., Thompson, D. T.: J. Chem. Soc. (London) Ser. A 190 (1968).

W. Strohmeier

54) Cramer, R.: J. Am. Chem. Soc. *88*, 2272 (1966).
55) Strohmeier, W., Rehder-Stirnweiß, W.: J. Organometal. Chem. *26*, C 22 (1971).
56) – – J. Organometal. Chem. *19*, 417 (1969).
57) – – J. Organometal. Chem. *22*, C 27 (1970).
58) Hudson, B. B., Taylor, P. C., Webster, D. E., Wells, P. B.: Discussions Faraday Soc. *46*, 37 (1968).
59) Hussey, A. S., Takeuchi, Y.: J. Am. Chem. Soc. *91*, 672 (1969).
60) Strohmeier, W., Rehder-Stirnweiß, W.: Z. Naturforschung *24b*, 1219 (1969).
61) Hallmann, P. S., Mc Garvey, B. R., Wilkinson, G.: J. Chem. Soc. (London), Ser. A 3143 (1968).
62) Jardine, F. H., Osborn, J. A., Wilkinson, G., Young, J. F.: Chem. & Ind. (London) 560 (1965).
63) Osborn, J. A., Wilkinson, G., Young, J. F.: Chem. Commun. *17* (1965).

Eingegangen am 28. Januar 1971

Elektronik der Trägerkatalysatoren

Prof. Dr. Georg-Maria Schwab

Physikalisch-Chemisches Institut der Universität München

Inhalt

1. Einleitung

Unter "Elektronik" soll hier, vielleicht abweichend vom verbreiteten Sprachge-
brauch, nicht der Aufbau und die Funktion der elektronischen Schaltelemente
verstanden werden, mit denen Vorgänge in und an Trägerkatalysatoren beobachtet werden können. Hier geht es vielmehr um die *elektronischen Vorgänge
innerhalb der Trägerkatalysatoren selbst,* auf denen ihre verbesserte, bisweilen
auch verschlechterte Wirkung beruht. Trägerkatalysatoren sind nämlich feste
Systeme, in denen eine katalytisch wirksame feste Komponente, der Katalysator,
in elektrisch leitendem Kontakt mit einem katalytisch ganz oder vergleichsweise
unwirksamen zweiten Feststoff steht, eben dem Träger.

Es hat sich in den letzten Jahrzehnten herausgestellt, daß es eine Gruppe
von Reaktionen, im allgemeinen Redoxreaktionen gibt, deren katalytische Beeinflussung durch Festkörper systematisch verstanden werden kann, wenn man
als maßgebenden Parameter den inneren elektronischen Aufbau des Festkörpers
betrachtet und ihn beeinflußt. Es sei hier von vornherein betont, daß zwar Katalyse eine Oberflächeneigenschaft und der elektronische Aufbau eine Volumen-
eigenschaft sind, daß aber beide sicherlich unmittelbar zusammenhängen und
man davon ausgehen darf, daß

Elektronenmangel im Volumen auch unbesetzte Orbitale in der Oberfläche und

Elektronenüberschuß im Inneren besetzte Orbitale in der Oberfläche bedeuten [1].

Die erwähnte Gruppe von Reaktionen teilt sich wieder in solche, bei denen
im geschwindigkeitsbestimmenden Schritt die thermische Aktivierung einen
Übergang von Elektronen aus dem reagierenden Substrat in den Katalysator in-
volviert und die daher Donatorrektionen genannt werden, und in solche, bei denen der mit der Aktivierung verbundene Ladungsübergang in der umgekehrten
Richtung erfolgt — Akzeptorrektionen. *Donatorrektionen* sind, grob gesprochen,
solche, die Wasserstoff oder Kohlenmonoxid, also Reduktionsmittel mobili-
sieren, wie Hydrierungen, Dehydrierungen, Parawasserstoff-Umwandlung und
Deuteriumaustausch, *Akzeptorreaktionen* solche, in denen Sauerstoff oder andere Oxydationsmittel (H_2O_2) mobilisiert werden. Für beide Gruppen hat sich
gezeigt, daß für metallische Katalysatoren und für halbleitende Katalysatoren
(Isolatoren bieten sich einer messenden Betrachtung des elektronischen Aufbaus
nicht an) der Richtung nach dieselben Gesichtspunkte gelten in dem Sinne, daß
Donatorreaktionen am besten durch elektronenarme Metalle (d-Metalle) und
p-leitende Halbleiter katalysiert werden, Akzeptorreaktionen dagegen durch Metalle mit vielen Elektronen (s-Metalle und ihre Legierungen) oder durch n-Leiter.

Diese Erkenntnisse, zusammengefaßt als "der elektronische Faktor in der
Katalyse", ließen es neuerdings aussichtsreich erscheinen, nun auch das Studium
der auffallenden Erscheinungen bei der Trägerkatalyse vom elektronischen Standpunkt aus neu in Angriff zu nehmen.

Seit Jahrzehnten kennt man ja zwei Arten der Beeinflussung eines aktiven Katalysators durch die Natur eines Trägers:

1. Die *"strukturelle Verstärkung"*, darin bestehend, daß ein großoberflächiger Träger auch dem auf ihm aufgetragenen Katalysator eine große Oberfläche erteilt oder erhält (technischer Ammoniakkontakt).

2. Die *"synergetische Verstärkung"*, bei der der Träger durch ein energetisches Zusammenwirken mit dem Katalysator neue Effekte hervorbringt.

Die synergetische Verstärkung ist phänomenologisch durch das Auftreten einer wesentlich erhöhten Reaktionsgeschwindigkeit bei wesentlich herabgesetzter Aktivierungsenergie gekennzeichnet. Die Korrelation mit anderen Festkörpereigenschaften und damit ihr "Verständnis" stellt ein echtes und herausforderndes wissenschaftliches Problem dar, um so mehr als so gut wie alle technischen heterogenen Katalysatoren von diesem Typus sind.

Der Versuch, dieses Problem vom elektronischen Standpunkt aus anzugehen, knüpft an die beschriebenen Erfahrungen mit Einstoffkatalysatoren an. Wäre es nicht vorstellbar, ja naheliegend, daß die elektronischen Eigenschaften und damit die katalytischen Eigenschaften eines "getragenen" Katalysators bei leitendem Kontakt durch die elektronischen Eigenschaften des Trägers beeinflußt werden? So muß eine kleine Menge eines Metalles, die auf der Oberfläche einer großen Menge eines halbleitenden Trägers verteilt ist, ihre Fermi-Grenze dem Fermi-Niveau des Halbleiters notwendig anpassen, und von der Lage der Fermi-Grenze innerhalb der Leitfähigkeitzone des Metalles sollte nicht nur dessen Leitfähigkeit, sondern nach Obigem auch dessen katalytische Wirksamkeit abhängen.

Dies sind die in der Technik so häufigen Fälle von Metallen auf halbleitenden Trägern, deren Brauchbarkeit jetzt einem neuen Verständnis zugeführt wird. Ein Beweis für eine solche Auffassung kann erbracht werden, wenn es gelingt, die katalytische Wirksamkeit des Metalles dadurch zu verändern, daß man das Fermi-Niveau des Trägers bewußt verändert. Dieses ist bei Halbleitern gut möglich durch die Methode des Dotierens: Zusatz von Kationen höherer Ladung zum Grundmaterial erzeugt Elektronen und vermindert die Defektelektronen, erhöht also das Fermi-Niveau, während Zusatz von niederwertigen Kationen das Umgekehrte bewirkt. Man muß also Metallkontakte auf dotierten Halbleiter-Trägern untersuchen.

Verwickelter liegen die Verhältnisse in dem erst in den letzten Jahren systematisch untersuchten umgekehrten Falle, den *"inversen Trägerkatalysatoren"*, wo ein halbleitender Katalysator in dünner Schicht oder in kleinen Partikeln mit einem metallischen Träger in Berührung steht. Zwar sind hier von vornherein stärkere Effekte zu erwarten, was sich auch bestätigt hat, weil Metalle im allgemeinen mehrere Zehnerpotenzen höhere Elektronenkonzentrationen aufweisen, als Halbleiter, jedoch ist die Theorie weniger einfach [2-7].

Die Aktivierungsenergie einer von einem Halbleiter katalysierten Redoxreaktion hängt nicht einfach von der Lage des Fermi-Niveaus ab, sondern eher von dem Abstand dieses Niveaus von den Bandkanten. Für Donatorreaktionen ist sie proportional dem Abstand von der Unterkante des Leitbandes, für Akzeptorreaktionen dem Abstand von der Oberkante des Grundbandes. Für die Theorie, auf die wir hier nicht im einzelnen eingehen, sei auf [42] und auf den Bericht von F. Steinbach verwiesen[a]. Nun werden die Bandkanten durch die Berührung mit dem metallischen Träger verbogen: Wenn, wie gewöhnlich, die Austrittsarbeit des Metalles kleiner ist als die des Halbleiters, sein Fermi-Niveau also vor dem Kontakt höher, so tritt nach dem Kontakt Elektronenladung in den Halbleiter über, wodurch die Bandkanten nach unten verbogen werden, das Leitband also in einer dünnen Randschicht sich dem Fermi-Niveau nähert, was günstig für Akzeptorreaktionen und ungünstig für Donatorreaktionen ist. Da es sich hier um eine Verbiegung der Bänder, d.h. eine Abstandsveränderung nur innerhalb einer Randschicht von der Größenordnung 10^{-6} bis 10^{-5} cm Dicke handelt, haben wir eine Möglichkeit der Prüfung der Theorie in der Veränderung der Schichtdicke des auf dem Träger ausgebreiteten Katalysators bzw. seiner Korngröße. Daneben besteht natürlich die Möglichkeit, die Elektronenkonzentration und Fermi-Energie des tragenden Metalls zu verändern. In wenigen Fällen [8] wurden auch Mischkatalysatoren aus zwei verschieden dotierten Phasen desselben Halbleiterkatalysators untersucht; die Überlegungen, die hier anzuwenden sind, sind den obigen analog. Ferner wird es notwendig sein, selbst wenn es gelingt, auf diese Weise die katalytische Wirksamkeit zu beeinflussen, daß ein Beweis dafür geführt wird, daß dies tatsächlich auf dem Weg über eine elektronische Beeinflussung geschehen ist.

Im Vorstehenden sind bereits abwechselnd die katalytische Wirksamkeit und die Aktivierungsenergie der katalytischen Reaktion erwähnt worden. In der Tat werden in den unten abgehandelten Beispielen im wesentlichen Änderungen der Aktivierungsenergie (wahren oder scheinbaren, je nach der Kinetik der Reaktion) besprochen werden. Die Rechtfertigung für ein solches Vorgehen liegt in dem Kompensationseffekt (Theta-Regel), also in der Tatsache, daß allgemein in einer Gruppe vergleichbarer Reaktionen (bei uns gleicher Reaktionen an Kontakten aus gleichen Katalysatoren mit verschiedenen Trägern) der logarithmische Häufigkeitsfaktor und die Aktivierungsenergie linear miteinander verknüpft sind. Dies hat zur Folge, daß unterhalb einer gewissen "isokatalytischen" Temperatur der Katalysator mit der kleineren Aktivierungsenergie stets auch der wirksamere ist. Da aber der Häufigkeitsfaktor nicht allein von der Aktivierungs-Entropie, sondern auch von der Größe der wirksamen Oberfläche abhängt, die bei verschiedener Trägerdotierung durchaus verschieden sein kann, ist es sicherer, einen Parameter zu betrachten, der unmittelbar mit der inneren Natur der Reaktion verknüpft ist, also die Aktivierungsenergie.

---

a) Vgl. diesen Band, S. 117.

Im folgenden sei zunächst in einer Tabelle das bisher vorliegende bzw. dem Berichterstatter bekannt gewordene Material zusammengestellt. Es sind dabei nicht beliebige technisch erprobte Systeme aufgenommen worden, sondern nur solche, in denen gezielt nach Effekten der geschilderten Art gesucht worden ist. In der Mehrzahl handelt es sich um "normale" Trägerkatalysatoren, Metalle auf Halbleitern, in einigen wenigen, aber gut charakterisierten Fällen um "inverse" Trägerkatalysatoren und nur in zwei Fällen um Halbleiter auf Halbleitern.

2. Normale Trägerkatalysatoren

Der erste planmäßige Testversuch, in dem die elektronische Natur der Trägerwirkung normaler Art geprüft wurde, ist in [12] beschrieben und in [11,14] ausgebaut worden. Es ist festgestellt worden, daß die Legierungen der Zusammensetzung Cu_2Mg und $CuMg_2$ bei der Dehydrierung der Ameisensäure ein ihrem Leitfähigkeitsverhalten nicht entsprechendes katalytisches Verhalten zeigen. Dies konnte darauf zurückgeführt werden, daß das Kupfer nach der Ameisensäure-Behandlung nicht mit Magnesium legiert, sondern im Magnesiumoxyd fein verteilt vorlag. Dasselbe gilt für Silber in der Legierung AgMg. Messungen der Aktivierungsenergie an diesen Metall-Oxydsystemen (Fälle 7, 13, 14 der Tabelle) zeigten, daß diese durch Anwesenheit des Trägers erheblich herabgesetzt war.

Daraufhin wurde eine Versuchsreihe mit dotierten Trägern angesetzt [11,14]. Gamma-Aluminiumoxid, das im Handel als n-Leiter vorliegt, wurde mit Germaniumdioxid oder Titandioxid in Richtung steigender, mit Nickeloxid und Berylliumoxid in Richtung fallender n-Leitung dotiert. Auf solche Tabletten wurden die Metalle Nickel, Kobalt und Silber in dünner Schicht aufgedampft und die Aktivierungsenergie der Ameisensäurespaltung gemessen. Die Werte der Tabelle (Fälle 7 und 8 sind Grenzwerte; zwischen ihnen ändert sich die Aktivierungsenergie systematisch mit der Dotierung des Trägers, wobei die Reaktion vom Donatortyp ist, wie aus Untersuchungen mit reinen Metallen bekannt ist) ergeben, daß die Elektronenkonzentration des Metalles um so geringer wird, je mehr sich der Träger dem p-Typ nähert.

Gleichzeitig wurden Änderungen der optischen Reflexionskante des Metalls sowie der auf die Feldstärke 0 extrapolierten Suszeptibilität gefunden [3,14], was für eine elektronische Beeinflussung des Nickels spricht. Für Kobalt und in geringerem Maße für Silber kann dann eine solche ebenfalls angenommen werden.

Es ist natürlich wünschenswert, zu prüfen, ob eine solche Beeinflussung auch noch eintritt, wenn der Katalysator und der halbleitende "Träger" nicht in Schichten übereinander liegen, sondern in der Art eines technischen Mischkatalysators als *Pulvermischung* in Kontakt miteinander stehen. Diesem Ziel dienen die Versuche mit der hydrierenden Wirkung des Nickels auf Äthylen (Fall 6 der Tabelle), wenn dieses Nickel mit n-leitendem Zinkoxid verschiedener Dotierungen vermischt ist [9,10] (siehe auch [3,7,11]). Hier konnte nicht nur gezeigt

Tabelle

Lfd. Nr.	Kata-lysator	Träger	Reaktion	Aktivierungsenergie Kcal/Mol				Variation	Anordnung	Lit.
				ohne Träger	p-dot.	undot.	n-dot.			
1	Ni	NiO	HCOOH $\longrightarrow$ $H_2 + CO_2$	24	3	8	10	Träger dotiert	Oberfl. red.	3)
2	Ni	Ge	desgl.	24	10	–	18	Träger dotiert	Schicht	3)
3	Ni	TiO_2	desgl.	24	13	–	25	Träger dotiert	Schicht	3)
4	Ni	Cr_2O_3	desgl	24	17,5	–	23	Träger dotiert	Schicht	3)
5	Ni	MgO	N_2H_4	–	16	14	11	Träger dotiert	Schicht	3)
6	Ni	ZnO	$C_2H_4 + H_2$	–	13,4	–	16,5	Träger dotiert	Mitfällg.	3,7,9-11)
7	Ni	Al_2O_3	HCOOH	27	7	20	24	Träger dotiert	aufdampfen	3,11-14)
8	Co	Al_2O_3	HOOOH	23	13	21	24	Träger dotiert	aufdampfen	3,11)
9	Fe	Al_2O_3	HCOOH	20	13	18	22	Träger dotiert	Mischoxid red.	7,11,15)
10	Fe	$Al_2O_3 + K_2O$	HCOOH	20	7–13	–	–	Träger dotiert	Mischoxid red.	7,11)
11	Fe	Fe_3O_4	HCOOH	20	–	–	30–40	Träger oxidiert	Mischoxid red.	7,11)
12	Fe	Al_2O_3	NH_3-synth.	13,5	13–15	10	10	K bzw. Ge	Mischoxid red.	15,16)
13	Ag	MgO	HCOOH	17	–	15	–	Verteilung	eingebettet	13)
14	Cu	MgO	HCOOH	20	–	15	–	Verteilung	eingebettet	13)
15	NiO	Ag	$2 CO + O_2$	16	–	16–46	–	El.-Konz.d.Ag, Schichtdicke	Schicht	17)
16	Fe_2O_3	Ag	$2 SO_2 + O_2$	31	24	–	7	El.-Konz.d.Ag	Schicht	18)
17	ZnO	Ag	CH_3OH-ox.	47	–	24	–	Verteilung	Mischung	19,20)
18	ZnO	Zn	Aether-bldg. HCOOH dehydrat.	28	–	18	–	Verteilung	Mischung	21)
19	ZnO	Zn	desgl.	5	–	25	–	Verteilung	Mischung	21)
20	Cu_2O	Cu	HCOOH dehydrat.	20	–	50	–	Verteilung	Mischung	21)
21	ZnO (Li)	ZnO (In)	$2 CO + O_2$	23 (16)	–	11	–	Dotierung	Mischung	8)
22	NiO (Li)	NiO (In)	$2 CO + O_2$	15 (8,1)	–	9,7	–	Dotierung	Mischung	8)

werden wie aus der Tabelle ersichtlich, daß die Hydrierung als Donatorreaktion bei Lithium-Dotierung des Zinkoxids rascher und mit kleinerer Aktivierungsenergie verläuft als bei entsprechender Gallium-Dotierung, sondern wegen der Konkurrenz von Adsorption und Reaktion in dieser Reaktion konnte zusätzlich gezeigt werden, daß die Adsorptionsenergie des Äthylens bei Lithium-Dotierung größer ist als bei Gallium-Dotierung. Dies hat sogar zur Folge, daß das Temperaturoptimum, das von der Konkurrenz der Äthylen-Desorption mit der Reaktion herrührt, bei Gallium-Dotierung wegen der verminderten Adsorptionsenergie ausbleibt.

Es lag nahe, zu prüfen, ob nicht in dem bekannten technischen Ammoniak-Synthese-Kontakt $Fe/Al_2O_3/K_2O$ ähnliche Zusammenhänge obwalten (Fälle 9 bis 12 der Tabelle) [7,11,15,16].

Zunächst wurde dies an der notorischen Donatorreaktion der Ameisensäure-Dehydrierung geprüft. Während reines Eisen eine Aktivierungsenergie von 20 kcal/Mol ergibt, wird diese durch den n-Leiter Al_2O_3 (kommerziell) auf 18 kcal/Mol erniedrigt, bei K-dotiertem Aluminiumoxid sogar auf 13, ja 7 kcal/Mol, während Dotierung des Aluminiumoxids mit Germaniumoxid in Richtung erhöhter n-Leitung 22 kcal/Mol ergab, der durch Teiloxydation des Eisens entstandene starke n-Leiter Fe_3O_4 sogar 30 bis 40 kcal/Mol. Bei der Ammoniaksynthese, die, wenn die Chemisorption des Stickstoffs als Nitrid-Anion geschwindigkeitsbestimmend ist, eine Akzeptorreaktion darstellt, sollte das Umgekehrte eintreten. In der Tat wird die Aktivierungsenergie des Eisen-Aluminiumoxid-Kontakts durch Kaliumzusatz von 10 auf 13 bis 15 kcal/Mol erhöht. Warum aber setzt man dieses Kalium zu, wenn es die Katalyse der Akzeptorrektion erschwert? Es konnte gezeigt werden, daß es auf dem Wege über das Fehlordnungsgleichgewicht im Aluminiumoxid dessen Selbstdiffusion vermindert und daher die Stabilität des Trägergerüstes während des Gebrauchs erhöht.

Es muß nun eine Anzahl von Fällen besprochen werden, die etwa zur gleichen Zeit in anderen Laboratorien studiert wurden. Zunächst handelt es sich um die Ameisensäurereaktion an Nickel auf verschieden dotierten Trägern (Fälle 1 bis 15 der Tabelle) [3], darunter Titanoxid, Chrom (III)-oxid und Nickeloxid. Dort wird auch über Nickel auf Germanium berichtet, auch werden an der Hydrazin-Spaltung mit Nickel (Fall 5) entsprechende Befunde erhoben. Dieselben Autoren [22] stellen auch fest, daß die Änderungen der Aktivierungsenergie gemischter Oxide für Ameisensäure beim Vorerhitzen sich in denjenigen eines von ihnen getragenen Nickels widerspiegeln, was die Wechselwirkung beider Phasen bekräftigt. Endlich sei auf eine Untersuchung [23] hingewiesen, in der Cyclohexan mit Platin auf Aluminiumoxid dehydriert wurde. In diesem Falle sind zwar keine Aktivierungsenergien gemessen worden, aber es wurde ein Abfall der Aktivität bei zunehmender n-Dotierung durch Zirkon, Tantaloxid und Wolfram-Trioxid des Trägers festgestellt. p-Dotierung des Trägers ist hier ohne Einfluß, wenigstens auf die Ausbeute.

3. Inverse Trägerkatalysatoren

Es ist schon in der Einleitung ausgesprochen worden, daß *a priori* größere Effekte zu erwarten wären, wenn der Katalysator ein Halbleiter und der Träger ein Metall wäre, denn dann muß nicht der große Unterschied in der Elektronenkonzentration beider Phasen durch ein entsprechendes Mengenverhältnis ausgeglichen werden, weil nunmehr der Träger erheblich elektronenreicher als der Katalysator ist. Der erste geplante Versuch in dieser Richtung erfolgte erst 1966, und es muß auffallend erscheinen, daß diese Möglichkeit nicht früher ausgenutzt wurde. In Fall 15 der Tabelle [17)] wurde Silberfolie elektrolytisch vernickelt und die Nickelschicht durch Sauerstoff oxidiert, so daß Nickeloxid-Schichten zwischen 75 Å und 15 000 Å entstanden, die sich als gleichmäßig und lückenlos erwiesen. Nun wurde die Aktivierungsenergie der Oxydation von Kohlenmonoxid an diesen Katalysatoren gemessen, also eine Donatorreaktion. Während trägerfreies Nickeloxid eine Aktivierungsenergie von 16 kcal/Mol aufweist, weichen die dünneren Schichten von 500 Å abwärts davon nach oben ab, und bei 124 Å werden sogar 24 kcal/Mol gemessen. Da die Reaktion nach kinetischen Ergebnissen eine Donatorreaktion ist, ist damit erstmals gezeigt, daß die Elektronen, die aus dem Metall in den n-Halbleiter immittiert werden, dort Defekt-Elektronen neutralisieren und damit das Grundband nach unten verbiegen, es in der Randzone (dünner als 500 Å) damit vom Ferminiveau entfernen, die Aktivierungsenergie erhöhen und den Katalysator verschlechtern. Es konnte sogar gezeigt werden, daß eine Verkleinerung der Austrittsarbeit des Silbers durch Antimon-Zusatz die Reaktion am Nickeloxid völlig verhindert (Fall 15 der Tabelle, s. auch [5, 23, 25, 26)]).

Wurde hier durch den metallischen Träger der Katalysator verschlechtert, so wurde im folgenden Beispiel (Fall 16) einer Akzeptorreaktion der Katalysator verbessert. Es handelt sich um die *Synthese des Schwefeltrioxids* aus Schwefeldioxid und Sauerstoff, wo nach früheren Befunden die Aktivierung des Sauerstoffs maßgebend ist [18)]. Hier wurde durch Reduktion entstandenes Silberpulver durch Eindampfen eines Sols von aus Eisencarbonyl gebildetem Fe_2O_3 mit einer dünnen Schicht des Oxids überzogen. Elektroosmotische Methoden sicherten eine lückenlose Schicht von 64 Å Dicke. Es wurde nun nicht die Schichtdicke variiert, sondern die Elektronenkonzentration des Silbers, indem diesem schon vor der Reduktion Palladium oder Quecksilber zur Erniedrigung oder Erhöhung der Elektronenkonzentration zugemischt worden war. Das Ergebnis hinsichtlich der Aktivierungsenergie (die Aktivität hatte den entgegengesetzten Verlauf) ist:

trägerfreies Fe_2O_3 : 31 kcal/Mol,

Palladium-Legierung als Träger: 24 kcal/Mol,

reines Silber als Träger: 13 kcal/Mol,

Quecksilberlegierung als Träger: 7 kcal/Mol.

Der Zuwachs an Aktivität beträgt bei quecksilberhaltigen Trägern extrapoliert das 800fache! Es ist nur schade, daß diese Reaktion ohnehin eine so gute Ausbeute besitzt, daß Steigerungen der Geschwindigkeit technisch nicht erforderlich sind! Immerhin ist bewiesen, daß die Herabbiegung der Bänder von der Elektronenkonzentration und damit der Austrittsarbeit des metallischen Trägers abhängt.

Auch hier war, wie bei den normalen Trägerkatalysatoren, das nächste Bestreben, von dem Schichtaufbau des Katalysators zu dem mehr konventionellen Pulvergemisch überzugehen. Das ist der Fall 17 der Tabelle [19,20,27]. Hier wurde Methanol mit Sauerstoff zur Oxidation über Silber, über Zinkoxid, über beide nacheinander und über eine innige Mischung beider geleitet und die Geschwindigkeit der Oxydation chromatographisch bestimmt. Während die einzelnen Komponenten eine katalytische Wirkung oberhalb 200 $^\circ$C entfalten und während ihre gleichzeitige aber getrennte Anwesenheit sich kaum von der Additivität entfernt, zeigt das Gemisch schon wenig über 100 $^\circ$C raschen Umsatz, ja, führt unter Umständen zur Explosion. Arbeiten von Steinbach (vgl. diesen Band, S. 117) zeigen, daß die Reaktion an Zinkoxid eine Akzeptorreaktion ist (Aktivierung des Sauerstoffs), und so ist anzunehmen, daß die Berührung mit Silber auch hier eine Verbiegung der Bänder im Zinkoxid hervorgebracht hat — wenn auch eine bifunktionelle Wirkung des Paares nicht auszuschließen ist. In unveröffentlichten Versuchen haben Schwab und Seemüller die einzelnen Aktivierungsenergien gemessen, wie sie in der Tabelle angegeben sind.

Hält man solche Befunde zusammen mit der Erfahrung, daß Metalle meist eine kleinere Austrittsarbeit haben als ihre eigenen Oxide, so liegt es nahe, anzunehmen, daß ein Metalloxid eine stark veränderte Aktivierungsenergie und Aktivität entfalten sollte, wenn es auf seinem eigenen Metall als Träger sitzt. Im Fall 18 bis 20 der Tabelle [21] wurde das versucht. Man ist hier in der Wahl der Reaktion stark eingeschränkt, weil stationär weder das Metall oxidiert noch das Oxid reduziert werden darf. Immerhin konnte für die Dehydratation der Ameisensäure und des Äthanols an Zinkoxid und für die Ameisensäure-Dehydratation an Cu_2O auf Kupfer eine starke Herabsetzung der Aktivierungsenergie gegenüber dem reinen Oxid gefunden werden (s. auch die Zusammenfassungen [5,24–26]).

4. Zwei Halbleiter

Nur in zwei Fällen wurden Misch- oder Trägerkatalysatoren untersucht, in denen sowohl der katalytisch wirksame Bestandteil wie der Trägerbestandteil ein Halbleiter ist. Der eine Fall [22] sind Mischungen Magnesiumoxid-Chrom (III)-oxid, wo die Aktivierungsenergie stark von der Vorerhitzung abhängt, weil Spinellbildung eintritt. Aus dem an sich sehr charakteristischen Verlauf mit der Vorerhitzung können vorläufig kaum Schlüsse über die elektronische Wechselwirkung gezogen werden. Günstiger liegen die Fälle 21 und 22 der Tabelle [8], wo zwei Proben desselben Oxids (Zinkoxid oder Nickeloxid) gemischt wurden, von denen die eine in Richtung auf höhere, die andere in Richtung auf niedrigere Elektronen- bzw. Defekelektronenkonzentration dotiert waren. Die Reaktion ist die Oxydation von Kohlenmonoxid. In der Tabelle sind unter "Aktivierungsenergie ohne Träger" die Aktivierungsenergie des Lithium-dotierten und (in Klammern) diejenige des Indium-dotierten Oxids angegeben, und die Mischung zeigt im Falle des Zinkoxids in jedem Falle eine Verminderung, im Falle des Nickeloxids eine starke Herabsetzung gegenüber einer der beiden Phasen, also immer eine Wechsel-wirkung.

5. Physikalische Messungen

Die Aussage, daß der elektronische Zustand eines Trägers von Einfluß auf die erforderliche Energie für den Durchtritt von Elektronen aus dem reagierenden Substrat durch die Katalysatorschicht in den Träger ist oder umgekehrt, ist gleichbedeutend mit der Aussage, daß die Schichten aus Katalysator und Träger *Gleichrichtereigenschaften* aufweisen müssen, oder umgekehrt ausgedrückt: Die aus chemischen Tatsachen erschlossenen Gleichrichtereigenschaften unserer Trägerkatalysatoren müssen durch physikalische Messungen geprüft werden, um die Schlüsse zu bestätigen. Das ist in der Untersuchung [28] erfolgt. Das in [17] als inverser Mischkatalysator angewandte Paar Silber-Nickeloxid wurde als Schicht-präparat ausgestaltet (auf Silber wurde Nickel elektrolytisch niedergeschlagen und oxydiert, eine raumladungsfreie Gegenelektrode aufgedampft und getempert). Die Charakteristik dieses Schichtpaketes zeigte, auf 1 cm^2 bezogen, 0,10 Ohm in Durchlaßrichtung (Silber negativ) und 200 Ohm in Sperrichtung.

Noch überzeugender ist aber folgende Messung [25,27,29]: Die spezifische Leitfähigkeit einer Halbleiter-Tablette sollte durch Zumischen von etwa 10% Silberpulver um etwa 10% zunehmen aus Gründen der Kurzschließung von Teilen der halbleitenden Strecke. Die tatsächliche Zunahme beträgt bei den n-Leitern Zinkoxid, Eisen(III)-oxid und Titanoxid 100% und mehr und sinkt mit steigender Temperatur, schon ein Anzeichen einer elektronischen Wechselwirkung. Noch charakteristischer ist das Verhalten der p-Leiter Nickeloxid und Kupfer-(I)-oxid. Hier sinkt durch Zusatz von 10% Silber die Leitfähigkeit um ähnliche Beträge, offenbar weil die in den Halbleiter immittierten Elektronen die Defekt-elektronen neutralisieren. Chromoxid zeigt beide Verhaltensweisen, je nachdem ob es durch Vorbehandlung mit Sauerstoff zum p-Halbleiter oder mit Wasserstoff zum n-Leiter geworden ist.

Bezeichnend ist auch das Verhalten von Thoriumdioxid. Es ist zwar ein n-Leiter, verhält sich aber in dem Silber-Leitfähigkeits-Test wie die p-Leiter, weil es als einziges Oxid eine kleinere Austrittsarbeit als Silber besitzt, also Elektronen in das Silber immittiert und nicht solche aufnimmt. Diese Versuche bestätigen unmittelbar die Annahme, daß in Randschichten bzw. in Körnern von der Korngröße, die der Dicke von Randschichten entspricht, ein Elektronenaustausch mit dem metallischen Träger erfolgt.

6. Schluß

Durch die in diesem Bericht angezogenen Versuche ist bewiesen worden, daß bei bestimmten Redoxreaktionen die Wechselwirkung zwischen Trägern und Katalysatoren durch die elektronischen Eigenschaften von Metallen und Halbleitern und durch die Randschichttheorie verständlich sind. Insbesondere zeigt es sich, daß die Änderungen der Aktivität, ausgedrückt durch die entgegengesetzt gerichteten Veränderungen der Aktivierungsenergie, bei inversen Mischkatalysatoren (Halbleiter auf Metallen) erheblich größer sind als im Falle der längst in der Industrie üblichen normalen Mischkatalysatoren. Dies ist durch das Verhältnis der absoluten Elektronenkonzentration verständlich. Es ist zu hoffen, daß diese Gesichtspunkte, die erst wenige Jahre alt sind, in günstigen Fällen auch zu *gezielter Herstellung von Katalysatoren* ausgenutzt werden können. Es muß dazu nur der elektronische Charakter der zu beschleunigenden Reaktion, der Leitungscharakter des Trägers, (der auch innerhalb gewisser Grenzen einstellbar ist) und die Konzentration der Leitungselektronen des Katalysators bekannt sein.

Danksagung. Der Deutschen Forschungsgemeinschaft und dem Fonds der Chemischen Industrie, die den größten Teil der hier berichteten Untersuchungen unterstützt haben, sage ich meinen verbindlichsten Dank.

7. Literatur

1) Schwab, G.-M.: Umschau *1965,* S. 766.
2) Hauffe, K.: Symposium Electronic Phenomena Chemisorpt. and Catalysis, Vol. 1. Moskau 1968, Berlin 1969.
3) Solymosi, F.: Catalyt. Rev. *1,* 233 (1967).
4) Schwab, G.-M.: Chem.-Ingr.-Tech. *39,* 1191 (1967).
5) – Beispiele Angew. Forsch. *1969.*
6) – Pont. Acad. Sci.Soc. V. *31,* 433 (1967).
7) – Dechema Monograph. *49,* 99 (1964).
8) Komatsu, W., *et al.:* Catalysis *15,* 43 (1969).
9) Schwab, G.-M., Mutzbauer, G.: Z. Physik. Chem. NF *32,* 367 (1962).
10) – – Naturwissenschaften *46,* 13 (1959).
11) – Angew. Chem. *73,* 399 (1961).
12) – Sitzungsber. Bayer. Akad. d. Wissenschaft. 1958, 6. Juni: Pure Apppl. Chem. *5,* 655 (1962).

G.-M. Schwab

13) –, Block, J., Müller, W., Schultze, D.: Naturwissenschaften *44*, 582 (1957).
14) –, Schultze, D., Block, J.: Angew. Chem. *71*, 101 (1958).
15) –, Putzar, R.: Chem. Ber. *92*, 2132 (1959).
16) – – Z. Physik. Chem. NF *31*, 342 (1962).
17) – Siegert, R.: Z. Physik. Chem. NF *50*, 191 (1966).
18) – Derleth, H.: Z. Physik. Chem. NF *53*, 1 (1967).
19) – Koller, K.: Am. Chem. Soc. *90*, 3078 (1968).
20) – Seemüller, H.: Bull. Roy. Soc. Belg., im Druck.
21) – Zettler, H.: Chimia *23*, 489 (1969).
22) Batta, I., Bánsági, T., Solymosy, F., Szabó, Z.G.: Acta Chem. Acad. Sci. Hung. *41*, 219 (1964).
23) Pfeil, W.: Z. Physik. Chem. (Leipzig) *243*, 52 (1970).
24) Schwab, G.-M.: Surface Sci. *13*, (1969) 198 (1969).
25) – Sitzungsber. der Bayer. Akadem. d. Wissenschaft *1968*, 1. März.
26) – Chimia *21*, 40 (1967).
27) – Kritikos, A.: Helv. Phys. Acta *41*, 1166 (1968).
28) – Bruncke, F.:   Z. Naturforsch. *24a*, 1266 (1969).
29) – Kritikos, A.: Naturwissenschaften *55*, 228 (1968).

Eingegangen am 2. Dezember 1970

Heterogeneous Photocatalysis

Priv.-Doz. Dr. Friedrich Steinbach

Institut für Physikalische Chemie der Universität Hamburg

Contents

1. Introduction

During a chemical reaction, chemical bonds between atoms are broken and new bonds are formed. The efficiency of a catalyst consists in its ability to favour the electron transitions which occur when bonds are dissolved and formed. A knowledge of the distribution and concentration of the electrons in the catalyst therefore plays a predominant role in the interpretation of catalytic processes [1-12]. To determine the importance of the surface bonds of a catalyst in the elementary step of a catalytic reaction, we have to change the *distribution of the electrons* over the quantum states of the bonds at the surface of the solid. The electron distribution must be altered where the reacting atoms may take notice of it, i.e. at the surface where the reaction occurs or very close to it, but not in the bulk.

By doping a semiconducting catalyst it is possible to alter the electron concentration, which leads of course to a redistribution of the electrons over the quantum states. However, it has been shown by measuring the work function that the Fermi potential at the surface of the solid is not influenced even by drastic shifts of the Fermi potential in the bulk [13-15].

Illumination of an insulating or a semiconducting catalyst with visible or ultraviolet light provides a unique opportunity to alter the distribution of the charge carriers over the quantum states in the catalyst surface. At the same time, a change occurs also in the *excitation state* of the bonds between the surface atoms of the solid and the molecules adsorbed at the surface. No other surface property is altered. The simplest arrangement involves continuous illumination of the catalyst during the reaction. A steady state of the electron distribution is generated, which is different from the thermal equilibrium in the dark and which leads to a change in catalytic activity.

The investigation of photocatalysis and photosorption enables us to study the role played by the excited states of the surface bonds of the catalyst lattice in the elementary step of the catalytic reaction.

After the light is switched off, the initial dark state of the surface, the thermal equilibrium, may be restored. This occurs only if the excitation of the bonds has been followed by reversible surface reactions and the reacting particles are still present in the gas phase. If the excitation of the bonds has been followed by slow or irreversible surface reactions, the return of the surface to the dark state may be very slow or even impossible. With the exception of these consecutive reactions, no other surface property is irreversibly changed by light absorption.

2. Direct Observation of the Interaction between Photons and Valence Electrons of the Semiconductor by Means of Photoelectron Emission

The absorbed photons transmit their energy $h\nu$ to *electrons in the lattice* of the solid near the surface. If the photon energy is sufficiently high, photoelectric emission is observed. This method of direct observation of the excited electrons enables information to be gained about the nature of the excited electrons and the initial states of the electron transitions. Thus, a complete analysis of the energy states of a semiconductor surface becomes possible.

Investigations carried out with Si, Ge, GaAs, InAs, InP and AlSb gave the following results [16-21]:

1. Only interband transitions have been observed, i.e. valence electrons are promoted from states in the valence band into the conductive band. In other words, semiconductor bonds are excited.

2. No surface states have been observed, although it would have been possible to observe photoemission out of surface states with a density as low as 1 state per 100 surface atoms.

3. Only direct transitions have been observed, i.e. transitions for which the *k*-vector is conserved.

We may conclude that semiconductor bonds in the surface and close beneath the surface are excited. The chemical action of these excited bonds is manifested in photocatalytic reactions.

3. Photoannealing Processes

The very simplest „photochemical surface reaction" − if we may call it a reaction − is the low-temperature photoannealing of *radiation defects* in semiconductor surfaces. During photoannealing, surface atoms change their places in the surface layer of the lattice of the solid, thereby restoring to the equilibrium order of the solid surface the state of disorder and nonequilibrium created in the surface by high energy irradiation. Observation of these processes shows the bond loosening due to light absorption in a very simple way, because only the atoms whose bonds are excited by photon absorption are able to change their places.

An example is the *photoannealing of Ge* [22]. Irradiation with electrons of 4 MeV at 77 °K converts n-conducting samples of Ge to p-type Ge. The p-conducting Ge is annealed at temperatures between 77 and 300 °K in the dark or in white light. Hall effect and conductivity measurements are used to determine sign and concentration of the mobile charge carriers. In the temperature range between 77 and 93 °K, illumination produces a decrease in conductivity. However, in the dark the original conductivity as it was before the disturbance of the surface is restored. In the temperature range between 93 and 153 °K, the reverse is observed: in the light the original conductivity is restored, in the dark the conductivity decreases to smaller values. At higher temperatures, the effects of light and dark annealing are again reversed. Only at temperatures higher than 213 °K do the effects of annealing move in the same direction in light as well as in dark. At all temperatures, only light with a photon energy higher than 0,32 eV is effective. The respective levels, therefore, are assumed to be situated in the lower half of the forbidden gap. The width of the forbidden gap in Ge is 0.73 eV [23]. The effects observed on the conductivity are attributed to alterations of the surface defects. Reversible conductivity changes are thought to be caused by electron excitation, while irreversible changes are attributed to a migration of surface atoms.

A similar investigation [24] has shown that the annealing process consists in the recombination of a Ge vacancy and an interstitial Ge atom. The rate of this recombination may be strongly influenced by light absorption.

Similarly, with $Pb(NO_3)_2$ *crystals* it has been shown that the annealing of surface defects generated by irrediation with γ-rays of a ^{60}Co source or by fast neutron irradation may be enhanced by illumination. The rate of annealing is a linear function of the frequency of the light [25].

The different ways light influences annealing processes all lead us to the same conclusion: light-induced electron excitations near and at the surface of the semiconductor are basically bond excitations, whether they appear as interband transitions or as transitions of electrons from impurity levels into bands. Electron excitation is a weakening of bonds; the return to the original state is a strengthening of bonds in the semiconductor. The quantitative correlation between photon energy and electron transition may not be as close in these experiments as in the photoelectron emission reported above. However, the loosening of bonds due to electron excitation is shown in a very simple and straightforward way: atoms can migrate from one location to another only when the bonds formed between them and their neighbours are temporarily loosened.

4. Photosorption and the Occupation of Electron Levels in the Solid

"Photocatalytic reactions with one reaction partner only" we may call photosorption processes at the surface of a solid. The investigation of photosorption processes is the first step towards an understanding of photocatalytic reactions. Because fewer parameters are involved than in photocatalytic reactions, a large number of investigations on photosorption have been carried out. For the same reason, photosorption will be discussed to some extent in this context.

By measuring a physical process related to the electron distribution in the surface, it may be possible to link the photosorption process at the surface with a certain electron transition in the surface layer of the catalyst lattice. Photoelectric emission, thermal afterglow, photoconductivity and Hall effect, changes in surface potential or electron spin resonance are such processes.

4.1. Photosorption and Related Processes on ZnO

Hall effect measurements enable the number of electrons in the conduction band of ZnO to be established during the chemisorption of oxygen. Though this investigation has been carried out in the dark only, the results are very important in this context [26]. Hall effect measurements may be carried out with polycrystalline sintered powders. They are not disturbed by grain boundaries, narrow connecting necks, pores and cavities, as are conductivity measurements. The use of powders increases the surface area of the samples, so that larger amounts of oxygen may be chemisorbed. In an earlier investigation [27] of the *CO oxidation on ZnO,* in which the same experimental method was applied, it was shown that one electron is removed from the conduction band during chemisorption of one oxygen atom. The electron is delivered back into the band during the reaction of the CO molecule with the chemisorbed oxygen atom, thus forming CO_2. The behaviour of electron concentration during oxygen sorption is now studied in greater detail at different temperatures. A small disc of Ga-doped ZnO powder sintered at 825 °C is used as a catalyst; the BET surface area is 0.85 m^2/g. The chemisorption is followed by measuring the manometric pressure of oxygen. Table 1 shows the number of electrons removed from the conduction band in comparison with the number of oxygen molecules chemisorbed. In the temperature range between 100 and 180 °C the ratio is 1, increasing to 2 in the range 230 to 250 °C. The increase may also be observed on a single ZnO sample: oxygen is chemisorbed at 160 °C; the numbers of molecules chemisorbed and electrons consumed are shown in Table 1. The sample together with the chemisorbed oxygen is then heated to 260 °C, where twice as many conduction electrons are removed, the ratio being now 2 conduction electrons per chemisorbed oxygen molecule.

From these results is it concluded that *oxygen is chemisorbed* as O_2^- in the temperature range 100 to 180 °C, at temperatures of 230 °C and above, oxygen is chemisorbed as O^-. Between 180 and 230 °C both oxygen species occur. A direct transition of the O_2^- species into the O^- species occurs when the temperature is raised. However, it is impossible to decide whether this transition occurs immediately at the surface or via a desorption readsorption process.

A closer correlation between the surface process and the charge carrier transition − change in the occupancy of the bond orbitals − in the surface of the solid is demonstrated by investigations of the *thermal afterglow* of polycrystalline ZnO samples and the influence of gas adsorption on the thermal glow cur-

Table 1. *Chemisorption of Oxygen on Zinc Oxide between* 100 ° *and* 250 °C

Temperature	Decrease in no. of electrons	Number of oxygen molecules chemisorbed	$\Delta e/O_2$
100°	4.9×10^{15}	5.0×10^{15}	1.0
150°	4.9	5.1	1.0
170°	5.1	4.8	1.1
180°	5.1	5.1	1.0
190°	6.8	5.5	1.2
200°	7.8	5.1	1.5
210°	8.6	5.2	1.7
220°	7.8	4.7	1.7
230°	10.6	5.3	2.0
240°	8.9	4.4	2.0
250°	9.8	4.7	2.1

ves [28]. The measurement of thermal glow curves is a direct method showing the existence of electronic trapping levels. Their energetic position may be computed directly from the glow curve [29-32]. The disadvantage of this as an experimental technique is that the surface contribution predominates over the bulk properties; this, however, is an advantage for the study of chemisorption and catalysis.

A semiconductor under a well-defined gas atmosphere or in vacuo is subjected to suitable optical excitation at a low temperature, usually that of liquid nitrogen, by ultraviolet illumination exceeding in energy the band gap of the semiconductor and filtered from infrared, which otherwise might induce quenching. A stationary state is produced with a specific occupancy of all electronic impurity levels which remains essentially unchanged when the ligth excitation is removed. Slow heating of the solid now leads to the release of the trapped electrons from the respective levels by thermal excitation at the temperature corresponding to the distance of a certain level from the bottom of the conduction band. The thermal liberation of the electrons from the trapping level may be determined by optical measurement of the thermal afterglow curve or by measurement of the thermally excited electron current. From the glow curves, the position and sometimes the occupancy of the individual energy levels may be determined.

When gases are adsorbed at the surface of the ZnO, the shape of the afterglow curves is influenced by the change in the occupancy of the level, some electrons of which interact with the substrate. By the adsorption of oxygen, air, water, hydrogen and hydrocarbons and the measurement of the activation energy of the respective level, a variety of impurity levels is established even in a high purity ZnO (less than 5 ppm total residual impurities) ranging from 0.12 eV to 1.2 eV below the bottom of the conduction band (Fig. 1). From this broad variety of impurity terms and from the pronounced influence of gas adsorption on the occupancy of the deep-lying levels, it is concluded that the correlation between sorption and occupancy of the shallow donor level at 0.12 eV, discussed in earlier interpretations, needs some revision. It seems more realistic to assume that the heats of adsorption on ZnO ranging from 8 to 18 kcal/mol may be correlated to this variety of impurity levels rather than to the single donor level lying only 2.7 – 3.5 kcal/mol below the band edge.

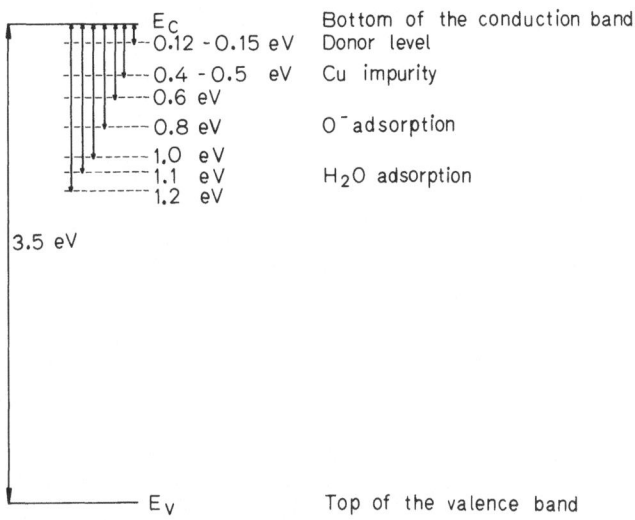

Fig. 1. Impurity levels in pure ZnO [28]

A more general correlation of oxygen sorption with processes near the surface of a ZnO crystal may be deduced from the measurement of the surface potential of a single crystal of ZnO during photosorption of oxygen [33]. A single crystal is mounted at a distance of 0.1 mm from a gold film evaporated on glass. The surface of the ZnO crystal and the gold film form a capacity. The surface potential of the single crystal is measured by an electrometer, which records the changes of potential due to oxygen photosorption.

Two processes may be distinguished from the changes in the surface potential due to illumination: a rapid rise and fall of the potential corresponding to short intervals of illumination and darkness. During intermittent illumination, the rapid process is accompanied by a slow overall rise in the surface potential. The slow rise is matched by a very slow decay in the dark. Both processes, the rapid and the slow one, are due to the behaviour of oxygen.

$$ZnO \leftrightharpoons Zn^{\cdot} + e' + 1/2\,O_2 \text{ (gas)} \tag{1}$$

$$O_2 \text{ (gas)} + e' = O_2^- \text{ (ads)} \tag{2}$$

$$O_2^- \text{ (ads)} + Zn^{\cdot} \rightarrow ZnO + O^X \text{ (ads)} \tag{3}$$

$$O^x \text{ (ads)} + e' \rightarrow O^- \text{ (ads)} \tag{4}$$

$$O^- \text{ (ads)} + Zn^{\cdot} \rightarrow ZnO \tag{5}$$

$$O_2^- \text{ (ads)} + e' \sim |e|^{\cdot} \rightarrow e' + O_2 \text{ (gas)} \tag{6}$$

$$ZnO + e' \sim |e|^{\cdot} \rightarrow e' + O_2 \text{ (gas)} + Zn^{\cdot} \tag{7}$$

ZnO, as a n-type semiconductor, has superfluous Zn atoms incorporated into the lattice as interstitial atoms. The interstitial Zn atoms are electron donors which, at room temperature, are dissociated completely into positive zinc ions and electrons (Eq. 1). This is the donor level 0.1 eV below the bottom of the conduction band, mentioned already in the discussion of the thermal afterglow curves. With respect to this level the following discussion seems to need some correction, since not the donor level but a deep-lying acceptor level may be involved in the reactions. However, this will not affect the general validity of the following interpretation.

In air or in oxygen atmosphere, oxygen is chemisorbed at the surface (Eq. 2). The chemisorbed oxygen may react with the interstitial zinc (Eqs. 3 to 5). This is an oxidation of the crystal surface. The ZnO lattice near the surface is thereby improved and the interstitial zinc atoms are consumed to form the perfect lattice. A chemical depletion layer is built up, depleted of interstitial zinc. Therefore, the positive space charge formed by the positively charged zinc donors to compensate for the negative surface charge of the chemisorbed oxygen species (left of Fig. 2) is moved away from the region near the surface into the lower depths of the lattice (rigth of Fig. 2), since there are no interstitial zinc donors in the immediate vicinity. This produces a drop in the surface potential.

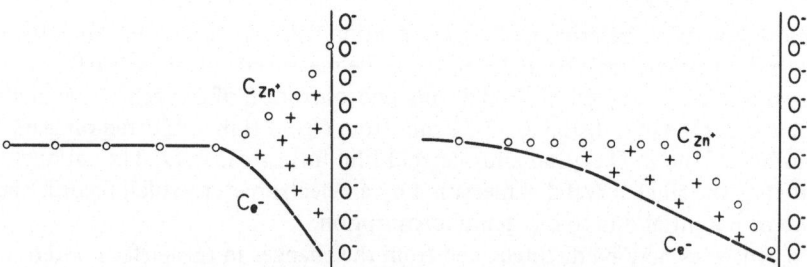

Fig. 2. Electrical and chemical depletion layer in ZnO [33]

During illumination of ZnO a photodesorption of oxygen occurs [34-40] (Eq. 6). Both photodesorption and readsorption in the dark (Eq. 2) are fast processes; they are responsible for the fast rise and fall of the surface potential. The slow rise and fall over several hours are due to a chemical reaction. In the light, a photochemical reduction of the surface takes place (Eq. 7), leading to a destruction of the chemical depletion layer, since interstitial zinc is produced. The positive space charge moves closer to the surface (left of Fig. 2) and a slow rise in the surface potential is observed. In the dark, a slow oxidation of the surface occurs (Eqs. 3 to 5). Thus again, the surface potential falls slowly over several hours.

Because of these slow processes, a ZnO surface "remembers" for some time whether and where it has been illuminated. The memory keeps the properties of the surface caused by illumination for some time after the illumination. This has been observed with respect to sorption and catalysis [41,42], surface potential, as just discussed, and conductivity. It was been observed also on other semiconductors, e.g. the exchange of oxygen isotopes and oxygen sorption on illuminated MgO [43]. The *memory effect* has been treated extensively by the electronic theory of catalysis [41,44,45].

The memory effect of the conductivity of ZnO makes *electrophotography* possible. Because of the close correlation between these phenomena and oxygen sorption, it is worth discussing these processes in some detail. Usually, the lifetime of photoelectrons in ZnO is about 10^{-5} s. In ZnO suited for electrophotography, the lifetime of photoelectrons is increased 10^9 fold by the presence of shallow trapping levels having a density of 10^{14} traps per cm^3 and situated 0.8 eV below the lower edge of the conduction band [46]. Fig. 3 depicts the band model of a ZnO grain in an electrophotographic layer. At the top of the figure the dark-adapted state of the grain is shown: oxygen is chemisorbed

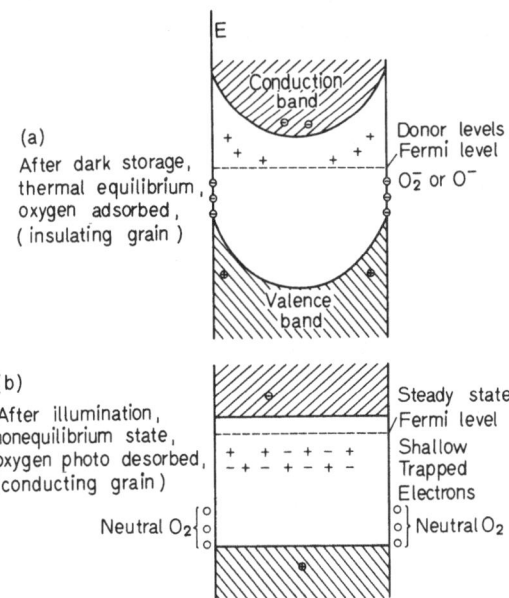

Fig. 3. Zinc oxide powder particle of 10^{-5} cm diameter. (a) Dark adapted, (b) light adapted (adapted from Ruppel, Gerritsen and Rose [46])

at the surface of the grain and the interior of the grain is depleted of electrons. Therefore, the mobile electrons left in the interior are unable to leave the grain because they are trapped by the high negative surface potential.

Electron hole pairs generated by photon absorption enable oxygen to desorb from the surface (bottom of Fig. 3). The oxygen desorption annihilates some of the holes, thereby decreasing the surface, so that electrons are now able to move from one ZnO grain to another. Thus, photoconductivity of the layer is produced. In the dark period which follows, the "photoconductivity" of the layer is preserved for some time due to the large number of shallow electron traps.

This model of an electrophotographic layer [46] has been supported by ESR investigations of electrophotographic ZnO [47], after the investigations of Sancier [41] had shown that ESR measurements can detect paramagnetic oxygen species (O^-) on the ZnO surface. When oxygen is chemisorbed in the dark at room temperature, O_2^- is at once formed at the surface. The signal of this oxygen species increases rapidly during further chemisorption of oxygen while a decrease in the conductivity of the ZnO is observed. One electron out of the conduction band is used for each O_2^- formed at the surface (see above [26]). If oxygen sorption is continued, a second signal is oberserved and this is attributed to O^-. At the end of the adsorption process a group of signals appears. These are characteristic for adsorbed molecular oxygen. Part of the chemisorbed oxygen migrates from the surface into the lattice and is incorporated as interstitial oxygen O_i^{2-}. The exchange of chemisorbed oxygen with the interior of the lattice has already been observed in experiments on the exchange of oxygen isotopes with a ZnO surface [49]. The O_i^{2-} does not produce any ESR signal; however, during illumination of the ZnO in vacuo it captures a defect electron, thereby being transformed to an O_i^- which is paramagnetic and produces a new ESR signal. If again oxygen is chemisorbed in the dark, new O_2^- species are generated at the surface by capture of electrons out of the conduction band. Thus, near the surface of the grains the depletion layer just described is built up. The high electric field of this depletion layer removes a defect electron from the O_i^- centres formed by previous illumination. The defect electron is consumed by the neutralization of the O_2^- and O^- species at the surface. During this process the O_i^- signal decreases slowly to zero (30 minutes at 300 °K). The fact that the width of the signals is independent of oxygen pressure shows that they are caused by an oxygen species in the interior of the lattice (O_i^-) and not by a surface species of oxygen. The interstitial oxygen may be removed completely by a five-minute vacuum treatment of the ZnO at 773 °K. The ZnO thus pretreated has lost the ability to produce the paramagnetic O_i^- centre in the light. Only after several hours' treatment in oxygen, may interstitial oxygen species be detected again.

A correlation of the amount of oxygen photosorbed with the content of interstitial zinc of the ZnO sample [50] and photoadsorption of oxygen at illuminated ZnO surfaces have been observed by several authors [51,52]. Both phenomena are interpreted by the Wolkenstein theory as being due to the different content of excess zinc in the different samples [53-55]. The photosorption of CO [56] and H_2O [57] on illuminated ZnO surfaces and the influence of the photosorbed gases on the conductivity of ZnO are interpreted in a way similar to the one just discussed with respect to oxygen sorption.

4.2. Photosorption on TiO$_2$

Experiments regarding photosorption on semiconductors other than ZnO have been carried out with the exception already mentioned [43] on TiO_2 only. After rapid oxygen adsorption in the dark coming to saturation coverage of the surface, rapid photoadsorption is again observed in a subsequent illumination period, followed by a further slow adsorption process [58]. The amount of oxygen adsorbed in the dark is not dependent on the oxygen pressure. Photosorption and dark adsorption of oxygen lead to a decrease in the conductivity of the TiO_2. The chemisorbed oxygen apparently withdraws electrons from the conduction band of the TiO_2 [59]. A more thorough investigation of the photosorption of oxygen on TiO_2 has been carried out by means of ESR techniques [60,61]. The TiO_2 sample outgassed at 10^{-5} torr at 500 °C for 1 hour gives rise to two ESR signals. One ($g = 2.002$) is attributed to a surface species of oxygen, the other ($g = 1.93$) to Ti^{3+} ions. If oxygen is admitted to the sample at 500 °C, a triplet signal is generated ("D", $g = 1.984$, 2.004 and 2.023) which is not influenced by high oxygen pressures, i.e. it is generated in the interior of the TiO_2 lattice. However, the intensity depends on the oxygen pressure during the treatment at 500 °C. If oxygen is admitted at room temperature to the TiO_2 sample a fourth signal ("O", $g = 2.009$) is generated. It is broadened by increasing oxygen pressure and must therefore be attributed to a surface species. Both signals are decreased during UV illumination at room temperature; simultaneously the oxygen pressure is increased. The intensity of signal D is decreased by illumination with light of wavelengths longer than 5250 Å; it is increased by light of wavelengths shorter than 5000 Å. In the dark period after the illumination, the intensity of the signal decreases.

An interpretation of the observed behaviour is possible if we consider the combined actions of photogenerated electron hole pairs and the capture and recombination of the charge carriers on surface centres produced by chemisorption or on lattice defects in the interior (Eqs. 8 to 13).

$$h\nu \rightarrow \ominus + \oplus \quad (e' \sim |e\,|\,\dot{}) \tag{8}$$

$$O_2(g) \rightleftarrows O_2(a) \tag{9}$$

$$O_2(a) + \ominus \longrightarrow O_2^- \tag{10}$$

127

F. Steinbach

$$O_2^- + \oplus \longrightarrow O_2(a) \tag{11}$$

$$\downarrow\uparrow \longrightarrow \uparrow + \ominus \tag{12}$$

$$\downarrow\uparrow + \oplus \longrightarrow \uparrow \tag{13}$$

Electron hole pairs generated by light absorption are recombined at surface centres produced by oxygen chemisorption (Eqs. 8 – 11). O_2^- should not be taken as a definite characterization, it is a surface species of oxygen generated by electron capture out of the lattice. Eqs. 10 and 11 are usually irreversible except at very high temperatures, when oxygen desorption takes place by the reverse of Eq. 10. Hence, the oxygen pressure influences the photoadsorption; however, it has no influence on photodesorption.

Therefore, photoadsorption is favoured by all processes which provide an efficient trapping of the holes, i.e. diminish the importance of Eq. 11. Photodesorption is observed at low oxygen pressures because the resulting low coverage of the surface by oxygen then limits the importance of Eq. 10.

Because signal D is independent of oxygen pressure, it is attributed to a *lattice defect*. It is postulated that signal D is produced when a non-paramagnetic lattice defect is ionized by transfer of the electron to the surface where it is trapped by chemisorbed oxygen. This accounts for the dependence of the intensity of signal D on the oxygen pressure during its formation at 500 °C. The formation of signal D is symbolized in Eqs. 12 and 13. Reaction 12 may be either a direct photostimulation of the defect or an indirect ionization due to the space charge in the depletion layer beneath the oxygen covered surface (Eq. 10).

Reaction 13 represents the production of a paramagnetic defect by capture of a photogenerated hole. The paramagnetic centres may be anihilated by the reverse of reaction 12 by capture either of photoelectrons from the conduction band or of electrons liberated during oxygen desorption.

These considerations lead us to the conclusion that the net photosorption depends on oxygen pressure, on the energies of the surface states of the chemisorbed species and on the participation of holes and electron traps in the reaction.

At low oxygen pressure, illumination produces photodesorption, as is shown by the increase of oxygen pressure and the decrease of the intensity of signal O. The simultaneous decrease of signal D is in accordance with a process whereby an electron is liberated from oxygen at the surface and is captured by a paramagnetic centre in the interior of the lattice (reverse of Eqs. 12 and 10). The subsequent behaviour when the ligth is turned off is interpreted as the readsorption of some oxygen. However, the capture of excess photoelectrons by the paramagnetic centres still predominates over the readsorption; therefore, a further decrease of signal D occurs.

128

At high oxygen pressure, photosorption of oxygen is observed and both signals (D and O) are increased. Apparently, signal D is produced by optical excitation of the electrons, which are then captured by oxygen chemisorbed at the surface (Eqs. 12 and 10). The wavelength dependence also supports the assumption of direct optical excitation. The defect level is situated in the band gap; it may be excited by wavelengths shorter than 5000 Å, and it may be occupied by an electron from the valence band in light of wavelenghts longer than 5250 Å. Therefore, the defect level must be situated about 2.4 eV below the lower edge of the conduction band. The band gap in TiO_2 is reported to have values between 3.0 and 3.67 eV [62].

We have seen from this review of photosorption processes how complicated even this simple case of a photoreaction with a solid surface may be. Photosorption studies make valuable contributions to the understanding of photocatalysis, since the atomic surface process and the charge carrier transition in the semiconductor may be connected more clearly than in a catalytic reaction. However, special care must be taken in generalizing results obtained in photosorption studies. In investigations of adsorption and photosorption, stable equilibrium or stationary states are studied which need not be identical with the transition states of a chemical reaction at the surface. Therefore, quite often the knowlege of photosorption will not be sufficient for an understanding of photocatalysis.

We must emphasize that the processes described in the language of the band model (electron excitation, photogeneration of electron hole pairs, migration and capture of charge carriers) are essentially processes whereby chemical bonds in the solid and at the surface of the solid are excited, broken and formed again.

5. Photocatalytic Reactions

In photocatalytic reactions the variations in the bond strength of the surface bonds due to photoexcitation are traced by the chemical action of the bonds, i.e. by changes of rate and by changes of reaction pathway of the reactions involving the bonds at the surface.

By *γ-irradiation*, charge carriers are generated in the bulk of Al_2O_3 and SiO_2. By diffusion of the charge carriers to the surface of the solids, the decomposition of methanol is enhanced [63].

Organic molecules adsorbed at the surface of SiO_2 are excited by light absorption and are enabled to undergo molecular rearrangement [64].

The catalytic activity of various semiconducting oxides and mixtures of oxides for the dehydrogenation of *aromatic hydrocarbons* is increased by ultraviolet irradiation [65]. The carbon monoxide oxidation photosensitized by ZnO has been studied by several authors (see below) [66-70].

An example of a decrease in the catalytic activity due to light absorption by the catalyst is furnished by ZnO. The catalytic oxidation of hydroquinone to quinone on aqueous suspensions of ZnO is greatly diminshed by illuminating the ZnO with light of wavelength 3650 Å [71].

Organic semiconductors, too, may be used as photocatalysts [72-74].

Electrode processes may be enhanced by illumination of one or both electrodes [75].

Paraffins and *olefins* are photooxidized on TiO_2 at room temperature, thus forming peroxides [76]. Consecutive radical reactions, usually occurring at higher temperatures in the dark, do not disturb the light reaction. The photocatalytic action of TiO_2 is well known from the industrial use of TiO_2 as white pigment. The resin varnish used to be destroyed by photocatalytic oxidation until it became possible to enclose the TiO_2 grains in a thin layer of silica. Hg vapour is photooxidized by TiO_2 to HgO at room temperature [77]. During the reaction H_2O is formed. The amount of HgO formed on the surface is equal to the number of basic OH surface groups of TiO_2.

5.1. Photoenhancement by Changes of Rate Constant

The photocatalytic activity of MgO, too, is considered to be caused by *surface hydroxyl groups.* This is shown in an investigation of the hydrogen-deuterium exchange at room temperature [78]. Monochromatic light was used in the range between 1800 and 4000 Å and with a band width of 96 Å. The reaction is treat-

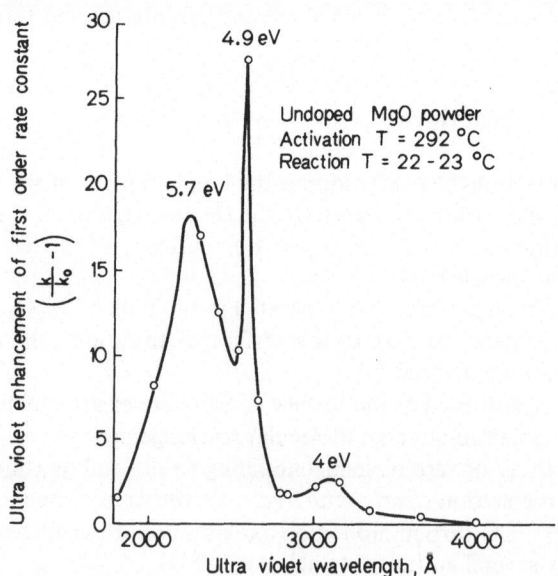

Fig. 4. Reaction rate spectrum for H_2-D_2 exchange on MgO (adapted from Harkins, Shang, and Leland [78])

ed as a first-order reaction. The photoenhancement is due to an increase in the rate constant, whereas the activation energy is the same in dark and light. With all wavelengths, after a linear increase in the rate constant with illumination time, a saturation value is reached after an illumination period of about 1 hour. These saturation values are used in the following results. Fig. 4 shows then enhancement of the rate constant at saturation in comparison with the dark value plotted against the wavelength. The resultant reaction rate spectrum indicates

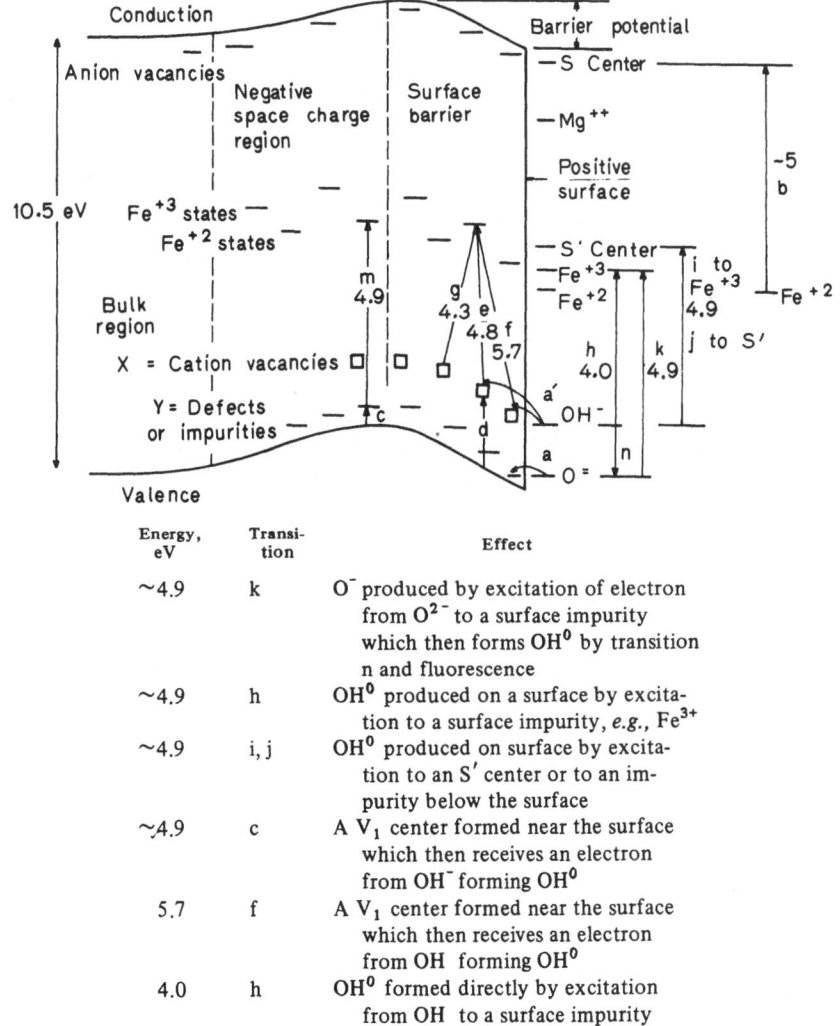

Energy, eV	Transition	Effect
~4.9	k	O^- produced by excitation of electron from O^{2-} to a surface impurity which then forms OH^0 by transition n and fluorescence
~4.9	h	OH^0 produced on a surface by excitation to a surface impurity, e.g., Fe^{3+}
~4.9	i, j	OH^0 produced on surface by excitation to an S' center or to an impurity below the surface
~4.9	c	A V_1 center formed near the surface which then receives an electron from OH^- forming OH^0
5.7	f	A V_1 center formed near the surface which then receives an electron from OH forming OH^0
4.0	h	OH^0 formed directly by excitation from OH to a surface impurity

Fig. 5. Band structure and possible transitions in MgO (adapted from Harkins, Shang and Leland [78])

3 peaks of photocatalytic response at approximately 4.0, 4.9 and 5.7 eV. The position of such peaks on the wavelength axis is a quantitative measure of the excitation energies effective in producing photogenerated active centres. The intensities of these peaks are an indication of the steady state concentration of photogenerated centres.

To understand these results, we must look at the band structure of the MgO surface (Fig. 5). In the band gap a number of lattice defects and surface states can be observed. The various transitions are shown by arrows. Vacuum pretreatment at 300 °C will cause oxygen chemisorbed as O^{2-} or O^{-} [79] to be removed from the surface or to migrate into the lattice, thereby annihilating an oxygen vacancy. Physically and chemically adsorbed water will be removed, thus occupying another O^{2-} vacancy in the surface:

$$2\,OH^- \longrightarrow H_2O + O^{2-}$$

Some of the hydroxyl groups remain at the surface. This is shown by the analysis of the HD formed after hydrogen–deuterium exchange at a surface which has not previous been in contact with hydrogen. Only by vacuum pretreatment at temperatures above 500 °C are the surface hydroxyl groups completely removed from the surface. The catalyst has then lost its activity. The iron impurity states alone are not responsible for the catalytic activity. Fe-doped MgO pretreated at 500 °C has a very low dark activity and the light influence is very small. The lifetime of S and S' centres is too short to be responsible for the catalytic activity. The ESR signal produced by these centres diminishes to zero over a period of time which is comparatively short, while the photocatalytic activity remains constant over several hours. On the right side of Fig. 5 are listed the transitions connected with the generation, annihilation and excitation of the surface hydroxyl terms. These must be considered as the active catalytic sites of the surface. The energies of these transitions are equal to those observed in the reaction rate spectrum of the preceding figure. The large number of possible transitions at 4.9 eV probably accounts for the high intensity of this peak in the spectrum.

From this we may draw conclusions regarding the reaction mechanism.

$$OH^- + h\nu \longrightarrow OH^0 + e^- \text{ (trap)} \quad \text{(direct)} \quad (14)$$

$$p^+ + OH^- \longrightarrow OH^0 \quad \text{(indirect)} \quad (15)$$

$$H_2 + OH^0 \rightleftharpoons HOH + H \quad (16)$$

$$D_2 + OH^0 \rightleftharpoons DOH + D \quad (17)$$

$$D + HOH \rightleftharpoons OH^0 + HD \quad (18)$$

$$H + DOH \rightleftharpoons OH^0 + HD \quad (19)$$

The activation of the OH^- site is achieved either by direct production of surface holes (Eq. 14) or by indirect production of holes and migration to the surface (Eq. 15). Both of these equations stand for the various transitions shown in Fig. 5. The hydroxyl level occupied by a defect electron reacts with H_2 and D_2 according to Eqs. 16 – 19. The authors point out that the hydroxyl group (OH^0) postulated in this study as an active site has never been detected in ESR studies [79,80]. This may be due to the fact that the MgO samples studied have been pretreated at temperatures higher than 300 °C. The reported results are remarkable because here, too, a deep lying trapping level in connection with a defect electron constitutes an active site, a finding which must be connected with the results obtained on ZnO [28,47,48] and TiO_2 [60,61,77].

The importance of the defect electron in surface processes on ZnO is also demonstrated in electrochemical investigations [81].

5.2. The Influence of Illumination on Activation Energy

While the photoenhancement of the hydrogen deuterium exchange at MgO has been found to be due solely to an increase in the rate constant, investigations of the photocatalyzed *carbon monoxide oxidation* with ZnO, NiO and Co_3O_4 as catalysts have shown that the photoenhancement in these cases is due to drastic changes in the apparent activation energy [82-88].

Whenever the rate constants of light and dark reaction respectively are of about the same order of magnitude, it should be possible to observe photoreaction and dark reaction simultaneously. Since the dark reaction is not detectable during the illumination of the catalyst, we conclude that the surface bonds excited by photon absorption interact with the non-excited bonds which might otherwise be responsible for the dark reaction. Therefore, in the interpretation of the following results the argument of bond excitation and bond-bond interaction in the catalyst surface is given greater emphasis.

In addition to the illumination of the catalyst surface, another simple method is used for the alteration of the electron concentration and the occupation of the bond orbitals in the semiconductor surface. This method is a modification of the *inverse mixed catalysts* introduced by Schwab [89-91]. The electron concentration and distribution upon the bond states is achieved 1. by putting the surface bonds into the potential of a boundary layer of a metal-semiconductor junction and 2. by illumination of the semiconductor-metal junction with ultraviolet light (photovoltaic effect).

When oxide grains 1000 Å thick are finely distributed on a metal film (Fig.6), the points of contact between the oxide grains and the supporting metal film build up an electric space charge layer (Schottky layer) which extends into the semiconducting grains. Because the diameter of the oxide grains is smaller than the thickness of an undisturbed space charge layer, the layer penetrates the grains and produces an alteration of electron concentration even at the opposite surface of the oxide grains. By illumination of the metal-semiconductor junction, with ultraviolet light passing through the metal, electron hole pairs are generated in the oxide. The potential drop at the metal semiconductor junction is partly decreased and a new concentration of charge carriers is produced in the oxide.

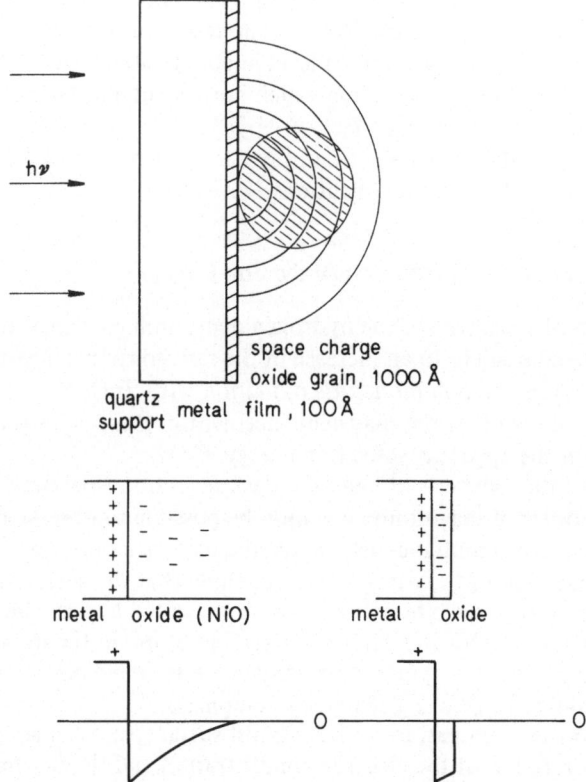

Fig. 6. Electric space charge layer in small oxide grains on metal films

The two charge carrier concentrations produced in the oxide either by contact potential or by photovoltaic effect must be distinguished from the charge carrier concentrations of the oxide alone. Therefore, the oxide grains supported on quartz without a metal film are investigated in dark as well as in light. These experimental arrangements enable four estimates of stationary electron concentration and electron distribution upon the quantum states of the surface bonds to be realized in one oxide: light and dark, with and without metal support. Four values pertaining to the catalytic activity are expected to correspond to the four stationary electron distributions in the orbitals of the surface bonds of the oxide grains. The catalytic activity of the surface is characterized by the activation energy and the pre-exponential factor of the carbon monoxide oxidation. Fig. 7 shows the results of experiments, according to the principles outlined above, with NiO, ZnO and Co_3O_4 on Ag films. The electron work func-

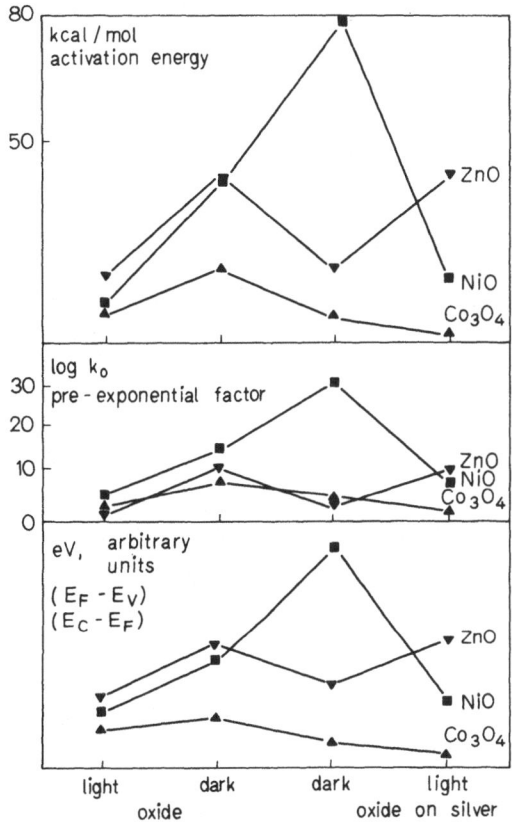

Fig. 7. Comparison of measured activation energies (top) and pre-exponential factors (centre) with the bond strength in the semiconductor surface expressed by the electron concentration in the surface (distance between Fermi potential and band edges, bottom)

tions of the p-conducting NiO and of the n-conducting ZnO are higher than the electron work function of Ag. The electron work function of the p-conducting Co_3O_4 is smaller than that of Ag. Thus, depletion layers and accumulation layers of electrons and defect electrons are investigated. Fig. 7 shows at the top the activation energies measured; the pre-exponential factors are shown below. There is a parallel influence on both parameters.

F. Steinbach

Fig. 8 demonstrates the curving of the bands in the three oxides. Because the diameter of the oxide grains is about ten times smaller than the thickness of a space charge layer, the charge carrier concentration in the grains corresponds to the arrows marked II. At the bottom of Fig. 7 is shown how the various experi-

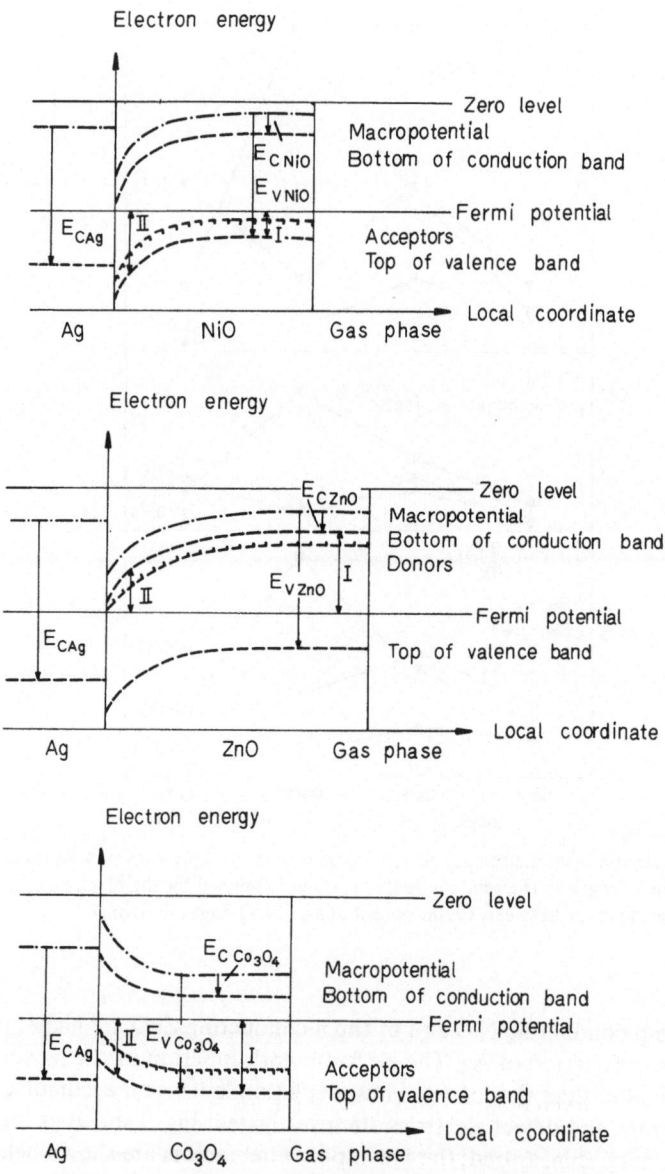

Fig. 8. Space charge at the silver semiconductor junction

mental conditions alter the distance between the Fermi potential and the band edge of the mobile charge carrier in the oxide grains. The proportionality between the activation energy (top) and this distance is clearly demonstrated.

According to an increase of $(E_F - E_V)$ in the depletion layer in NiO, the activation energy on NiO is raised by the influence of Ag support. The activation energy on ZnO on Ag support is decreased according to the decrease of $(E_C - E_F)$ in the accumulation layer in ZnO. The electron work function of Co_3O_4 is smaller than that of Ag, so that even in the dark, electrons flow into the silver, generating an accumulation layer of defect electrons. $(E_F - E_V)$ is decreased and so is the activation energy.

During permanent illumination of the semiconducting oxides without Ag support, a stationary state different from the thermal equilibrium is formed. It is characterized by the appearance of the quasi-Fermi potentials of electrons and holes shifted closer to the band edges.

Electron hole pairs generated in the oxide grains by illumination are separated by the electrostatic field of the contact potential of the metal-semiconductor junction, causing a net back flow of electrons from ZnO and NiO into Ag and from Ag into Co_3O_4. As the electron concentration in ZnO and NiO decreases, the curving of the bands in the space charge layer is decreased, too. As a consequence, the photoactivation energy decreases on p-conducting NiO and increases on n-conducting ZnO in comparison to the dark value. The electron concentration in Co_3O_4 increases and, accordingly, the curving of the bands is lessened as compared with the curving in the dark. However, under illumination the quasi-Fermi potential replaces the Fermi potential: the photoactivation energy on Co_3O_4 on Ag, where an accumulation layer is formed, is smaller thant that on Co_3O_4 without Ag support. For the same reason the photoactivation energy on NiO on Ag, where a depletion layer is formed, is higher than that on Ni without Ag support.

5.3. Correlation between Activation Energy of the Catalytic Reaction and Conductivity

We may conclude that the activation energy of the CO-oxidation is proportional to the distance between the Fermi potential and the band edge of the mobile charge carrier at the surface of the catalyst. This proportionality is demonstrated exactly in experiments on NiO in the dark [83].

In Fig. 9 (centre) the conductivity of NiO is plotted against the reciprocal temperature. The plot exhibits four different activation energies for four temperature ranges. The ratio of the activation energies of the conductivity is $4 : 2 : 0 : 1$. The same ratio $80 : 40 : 0 : 21$ is exhibited by the activation energies of the CO oxidation (top of Fig. 9). By an impurity level analysis the general dependence of the conductivity on $(E_F - E_V)$ is reduced to a dependence of the activation energy on the excitation energies $(E_A - E_V)$ of impurity levels of

137

the energy E_A. In Fig. 9 (bottom) the result is compared with the experimental measurements (top). In order to gain a clearer view, all ordinates have been reduced.

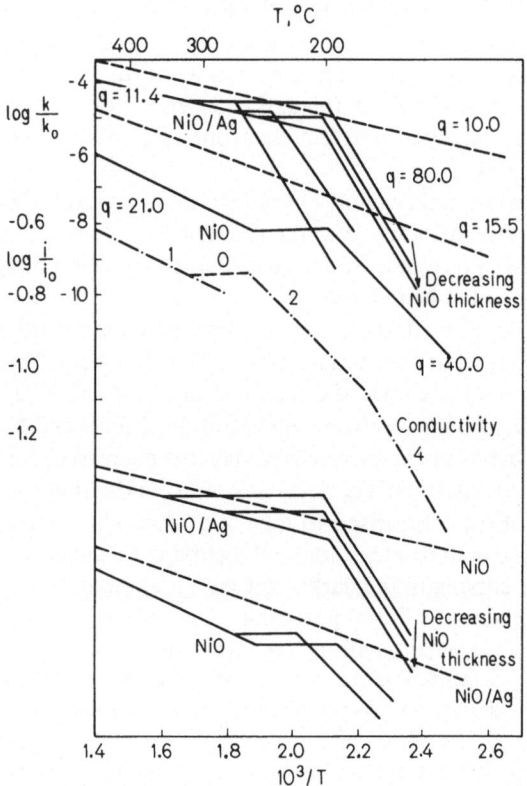

Fig. 9. Reduced Arrhenius plots (k/k_0 against $1/T$) of NiO and NiO/Ag. Comparison of measured activation energies of CO oxidation (top) with the conductivity of NiO (centre) and the Arrhenius plot obtained by impurity level analysis (bottom). Solid lines, dark; dotted lines, light; dash-dotted, conductivity

In NiO without support at low temperature at first the full excitation energy, then – with rising temperature – the half value of the excitation energy, and finally – as an horizontal line in the plots – exhaustion of the partly compensated acceptors is measured. Above the Néel temperature of NiO a second species of acceptors appears. It is characterized by a smaller excitation energy and appears after the exhaustion of the "low-temperature acceptors" with further increasing temperature. The conductivity plot (Fig. 9 centre) exhibits these four values; the catalytic reaction on NiO only three (40, 0, 21) and the reaction on NiO/Ag also the low-temperature value 80. The curves obtained by impurity level analysis (bottom of Fig. 9) demonstrate how successive poisoning of the impurity levels is achieved by the electrons of the silver support when the thickness of the layers of NiO grains is decreased. The experimental lines agree well with the values of the excitation energies obtained by impurity level analysis. The relation between the thickness of the NiO layer and the magnitude of the region of exhaustion is the same in experiment and theory. The correspondence of $(E_F - E_V)$ deduced from conductivity measurements and of the activation energies of the catalytic reaction is clear.

These results show that there is a direct proportionality of the activation energy of a catalytic reaction with the distance between Fermi potential and band edge; it may be expressed by the two empirical equations (20, 21), where a_n and a_p are factors specific for the reaction and the catalyst studied.

$$q = a_n (E_C - E_F) \qquad (20)$$

$$q = a_p (E_F - E_V) \qquad (21)$$

Investigation of a variety of oxide-silver combinations has thus provided a secure basis for the interpretation outlined above. A more refined set of experiments was then carried out by combining one oxide, NiO, with a series of supporting metals: silver, gold, palladium, platinum [88]. The work functions of these metals increase in this order. Corresponding to the decreasing differences between the work functions of the supporting metal and those of the oxide, there should be a decreasing influence on the catalytic activity of the oxide.

Fig. 10 shows the influence of the supporting metal on the activation energy and on the pre-exponential factor. Since the work function of NiO is higher than that of each supporting metal, a depletion layer of defect electrons is produced in the NiO in all experiments. The curving of the bands in the NiO grains, however, decreases from the pair NiO/Ag to the pair NiO/Pt. Accordingly, the increase of the activation energy of the CO-oxidation on NiO in the dark is highest when supported by Ag (the low-remperature value is doubled from 40 to 80 kcal/mole). It is less when supported by Au (from 21 to 39 kcal/mole, there is a 1.86-fold increase), and the smallest increase is produced by Pt (from 21 to 32 kcal/mole, there is a 1.53-fold increase). The high-temperature value of 21 kcal/mole is decreased to 11.4 by the silver support.

The contact potential is proportional to the difference of work functions in vacuo. The activation energy of the reaction depends on the distance between Fermi potential and band edge for each catalyst NiO/Ag, NiO/Au, NiO/Pd, NiO/Pt, as already described. Therefore, a linear correlation should result between activation energy and work function in vacuo of the supporting metal. In Fig. 10 the most reliable values of work function of the four metals are plotted, the plot forming a straight line. Likewise, the plot of the activation energy of the reaction on NiO on the corresponding metal support forms a straight line (with the exception of the value on NiO/Ag). Taking into account the simplifications of the interpretation outlined above, the experiments confirm the empirical relation between the activation energy of a catalytic reaction on the one hand and the distance between the Fermi potential and the band edge of the mobile charge carrier on the other.

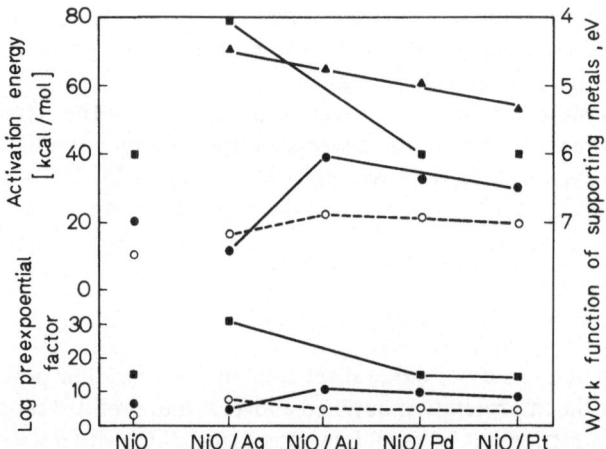

Fig. 10. Top: Comparison of activation energy on NiO and work function of supporting metal. Left scale: activation energy in kcal/mol. ■ Low temperature, ● high temperature values in the dark, ○ light. Right scale: work function of supporting metals in eV (▲). Bottom: Pre-exponential factor. The values on NiO without support are given on the left for comparison

The electron hole pairs generated in the NiO by illumination of the metal-semiconductor junction are separated by the electrostatic field of the electric double layer on the metal-semiconductor junction. The backflow of electrons from NiO into the support is proportional to the potential difference of the electrostatic field and thus to the work function of the supporting metal. Photoactivation energy, therefore, should be proportional to the potential difference and consequently to the work function of the supporting metal. Fig. 10 indicates that the experimental results confirm this for the supporting metals Au, Pd, Pt, whereas the photoactivation energy on Ni/Ag is an exception (perhaps because of the heavy poisoning of the impurity levels by the large number of electrons which immigrated in the dark). Furthermore, the difference between the activation energy in the dark and the photoactivation energy should decrease with increasing work function of the supporting metal, i.e. decreasing potential difference and photo-electromotive force at the metal-semiconductor junction. Fig. 10 also confirms this prediction.

Before going on, a short remark is necessary regarding the material to which the measured activation energy may be attributed. The oxide layer on the metal support is not coherent. The approximately 1000 Å thick oxide grains do not form a compact film. Regarding the oxide-metal catalyst, the question arises as to how the activation energy observed on a combination of metal and oxide is produced, and which parts of the surface of the catalyst are responsible. There are three possible interpretations:

1. The reaction may take place at the same time, but independently, on the oxide grains and on the metal surface not covered by oxide.

2. The reaction may take place on the oxide grains only. The electron concentration of the grains is changed by the formation of a space charge layer.

3. The reaction may take place on the metal only. The electron distribution of the metal may possibly be altered by the contact with the semiconducting grains.

From the experimental results, it can be deduced that only the second interpretation is valid [88].

By a similar reasoning it may be shown that the photoactivation energy observed on the oxide-metal layers is to be attributed to the electron distribution generated in the oxide by the photovoltaic effect [88].

5.4. The Influence of Wavelength and Light Intensity

In order to arrive at an interpretation of the relationship between activation energy and occupation of bond orbitals of the catalyst surface (energy bands), we investigated the influence of intensity of illumination and wavelength on the catalytic activity of the CO oxidation on ZnO [92].

There are two ways of varying the excitation of the bonds of the catalyst by illumination: by changing first, the wavelength of the exciting light and, secondly, the light intensity. Light of different quantum energies will excite different types of bonds, also electrons of one bond may be excited into different energy levels. Variations of light intensity will change only the number of bonds excited. In either case, one might expect an influence to be exerted on the catalytic activity. From Fig. 11 it is seen that both activation energy and pre-exponential factor are influenced in the same way. The activation energy decreases from the dark value of 23 kcal/mol, passing through the values of different wavelength ranges (filters) and intensities (metal gauze) to the smallest value, 15.8 kcal/mol, of the full arc.

Four results have to be pointed out:

1. The activation energy decreases with increasing light intensity I, following a parabolic law: $q_I = q_{dark} - \text{const } I^2$.

2. Accoring to the higher quantum energy, light of shorter wavelengths produces a stronger decrease of activation energy than light of longer wavelengths.

3. However, the efficiency of each wavelength is greatly diminished by simultaneous irradation with light of another wavelength. We may conclude that simultaneous illumination by quantums of different energy leads to a mutual interference of the effects of quantums of one energy.

4. In the light from several filters, horizontal parts with activation energy zero are observed in the Arrhenius plot. We must emphasize that this is observed in the same temperature range where, in the dark, the activation energy 23 kcal/mol is observed.

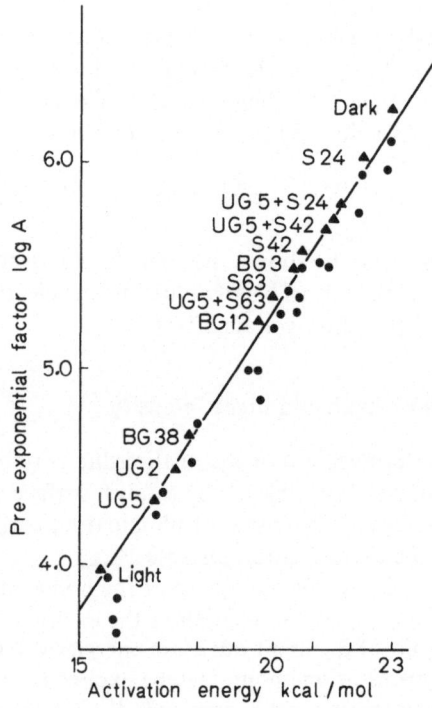

Fig. 11. Linear correlation between activation energy and pre-exponential factor for different wavelengths (filters UG and BG) and intensities (metal gauze sieves S) of the CO oxidation on ZnO

How may we understand these four experimental results? Let me say first how they cannot be interpreted. We cannot understand these four results by assuming that the reaction observed with each light intensity or each wavelength is composed of the sum of the reactions occurring simultaneously and singly on surface sites excited to different levels or not excited. If this assumption were made, the dependence on light intensity would follow a linear, not a parabolic law. It would follow that quantums of one energy cannot interfere with quantums of another energy. Furthermore, the appearance of horizontal parts in the Arrhenius plot with activation energy zero were not possible in the light, because the dark reaction occuring simultaneously on the sites not excited by photon absorption ought to be observed in the light, too.

Therefore, we believe that the surface of the ZnO may be considered as an extended resonance system of surface bonds. Now, an excitation of single bonds is not possible, but only an excitation of the system as a whole. Let us look at the mechanism of the reaction:

$$M - \overline{\underline{O}}| \; + \; |C \equiv O| \rightarrow \overset{\curvearrowright}{M} \overset{\curvearrowright}{-} \underset{\curvearrowright}{\overline{\underline{O}}}|^{\curvearrowleft}|C \equiv O| \rightarrow M| \; |\overset{\ominus}{\overline{O}} - \overset{\oplus}{C} \equiv O| \rightarrow M| \; + \; \overset{<}{O} = C = O\overset{>}{} \quad (22)$$

$$M| \; + \; |O \mathrel{\vdots} O| \rightarrow M|_{\underset{\cup}{\overline{O}} - \overline{\underline{O}}|} \rightarrow M - \overline{\underline{O}} - \overline{\underline{O}}| \xrightarrow{\;+M|\;} M - \overline{\underline{O}}| \; + \; M - \overline{\underline{O}}| \quad (23)$$

Eqs. 22 and 23 show the electron transitions necessary during the formation of CO_2 from a CO molecule and an oxygen atom of the surface. The reaction according to Eq. 22 is rate-controlling. In the rate-determining process a zinc oxygen bond in the surface is dissolved. It is restored in the fast reaction according to Eq. 23. The rate-controlling process consists of two steps (Eq. 24): first, a reversible electron shift, and secondly, an irreversible shift of the cores into the new positions.

$$
\begin{array}{cc}
Zn - \overline{\underline{O}}| & |C \equiv O| \\
& \\
& \updownarrow \text{ reversible} \\
& \\
Zn \cdots \overline{\underline{O}} \cdots\cdots\cdots\cdots C \equiv O| & \quad (24)\\
& \\
& \downarrow \text{ irreversible} \\
& \\
Zn| & O = C = O
\end{array}
$$

Oxygen moves irreversible out of the potential valley of a zinc oxygen bond into the potential valley of an oxygen carbon bond of CO_2. The activation energy of this step is controlled by the strength of the zinc oxygen bond. The bond may be described by a resonance of an ionic, a covalent and a nonbonding structure. The linear combination of these three main structures gives the actual bond strength:

$$\Psi_{ZnO} = a \, \Psi_{Zn^{2+} O^{2-}} + b \, \Psi_{Zn^{+} - O^{-}} + c \, \Psi_{Zn| \overline{\underline{O}} |} \quad (25)$$

When light is absorbed by the electrons of the bond, the bond is excited and weakened. The coefficient c of the nonbonding form increases in value while the coefficients a and b decrease: the potential valley of the zinc oxygen bond now becomes more shallow (Fig. 12) and the oxygen transition is favoured. Since we are dealing with a model based on resonance, an excited bond does not remain isolated from the neighbouring bonds, but the excitation energy is distributed over the whole system.

From this description of the semiconductor bonds it is possible to understand how there is a continuous decrease of the bond strength from the dark value to the value of the full arc by a continuous variation of the coefficients a, b and c. Therefore, a continuous change of the activation energy from the dark value to the value of the full arc may be observed. Interaction of the photons is made possible by a disturbance of the resonance of the bonds. This explains, especially, why it is impossible to observe a dark reaction in the stationary state of illumination.

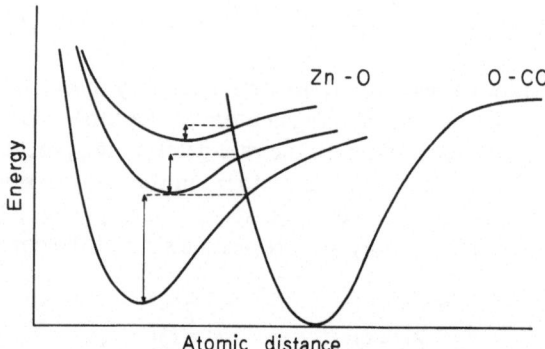

Fig. 12. Influence of bond strength of the ZnO bond on activation energy

In good agreement with this interpretation is the observation that the catalytic oxidation of hydroquinone on ZnO, well confirmed in the dark, is not observed in the light [71].

The promotion of an electron or defect electron to the conduction band or valence band, respectively, is only a part of the whole reaction. This excitation is identical with the destruction of a bond. In the band model only that part of the bond destruction is described which is connected with electron movement; the shift of the cores from the energy valleys is not taken into account. The activation energies of the conductivity and of the chemical reaction are proportional but not identical.

Now we can understand the results obtained with the oxide metal combinations as catalysts. Ag support increases the number of electrons in the lattice of NiO and ZnO and hence the number of resonance structures. In NiO, as a p-conductor containing fewer electrons than bonding orbital states, the number of bonding structures increases and the bond becomes stronger. The activation energy of a reaction that includes bond destruction, as demonstrated in Eq. 22, increases. In the light, electrons flow back into the silver. The total number of bonding structures decreases; moreover, the remaining structures are excited and the bonds are loosened. Bond destruction requires less energy, and so the activation energy of a reaction according to Eq. 22 is decreased. The number of electrons in the illuminated NiO/Ag is still greater than that in the illuminated NiO without support; in the former the bond remains more stable. This explains the higher photoactivation energy on NiO/Ag compared with NiO alone. The results obtained with the other supporting metals of NiO may be understood in the same way. By the same arguments the bond strength at the surface of the other two oxides can be shown to be reduced to the occupation of orbitals, i.e. to the electron concentration in and near the surface.

The simultaneously occurring changes of the pre-exponential factor are easy to understand as being due to the establishment of oxygen equilibrium in the oxide surface according to the strength of the semiconductor bonds.

Table 2. *Kinetics on NiO on metals*

	NiO	NiO/Ag	NiO/Au	NiO/Pd	NiO/Pt
$-dp_{CO}/dt =$	kp_{CO}	$k\dfrac{p_{CO}}{p_{CO_2}}$	$k\dfrac{p_{CO}^*}{p_{CO_2}}$	$k\,p_{CO}$	$k\,p_{CO}$

* In the light and at temperatures above 430 °C in the dark: $k\,p_{CO}$

This interpretation of the results by attributing them to different bond strengths is confirmed by the behaviour of CO_2 inhibition (Table 2). CO_2 inhibition occurs when the bond strengths is high, i.e. weak participation of the non-bonding structure in the linear combination of the bond. Consequently, CO_2 inhibition does not appear on NiO without support. It does appear on NiO/Ag in the dark and in the light, due to the numerous electrons which have immigrated from the Ag. On NiO/Au it no longer appears in the light, nor at higher temperatures in the dark, because of the smaller number of electrons that have immigrated. On NiO/Pd and NiO/Pt it does not appear at all.

5.5. Isopropanol Decomposition on ZnO, a Reaction Without the Transition of Oxygen

The relation between the excitation state of the surface bonds and the activation energy, discussed so far, is easy to understand when the rate-determining step of the reaction considered involves the dissociation of the more or less excited surface bond. In order to investigate the importance of the excitation state of the surface bond in a reaction without oxygen transition, the decomposition of isopropyl alcohol vapour on a ZnO catalyst in the dark and in ultraviolet light has been studied [93-95].

In the temperature range between 180 and 340 °C in UV light and in the dark, isopropanol is decomposed in two parallel reactions to acetone and H_2, and to propylene and H_2O, respectively.

F. Steinbach

$$
\mathrm{CH_3CHOHCH_3} \quad
\begin{array}{l}
\nearrow \quad \mathrm{CH_3COCH_3 + H_2} \\[2mm]
\searrow \quad \mathrm{CH_2{=}CHCH_3 + H_2O}
\end{array}
\qquad (26)
$$

dark and light
ZnO

Within the pressure range 2 to 30 Torr isopropanol, the dehydrogenation is a first-order reaction with strong inhibition by water. The simultaneously occurring dehydration is a zero-order reaction (Eqs. 27 to 29). Illumination was found to produce no change in the kinetics.

$$
\frac{dH_2O}{dt} = k_1 \qquad (27)
$$

$$
\frac{dA}{dt} = k_2 \frac{I}{H_2O} \qquad (28)
$$

$$
-\frac{dI}{dt} = \frac{dA}{dt} + \frac{dH_2O}{dt} = \frac{k_2}{k_1}\frac{I}{t} + k_1 \qquad (29)
$$

Fig. 13 shows the activation energies. They depend upon the age of the catalyst. The acitvation energy of the dehydrogenation is higher than that of the dehydration. UV light and increasing age of the catalyst diminish both activation energies. The light influence decreases with increasing age of the catalyst and disappears completely on the 20-day-old layers. We may recall that no oxygen is present in the reacting gas phase. Therefore, a continuous oxygen desorption takes place in the ZnO surface.

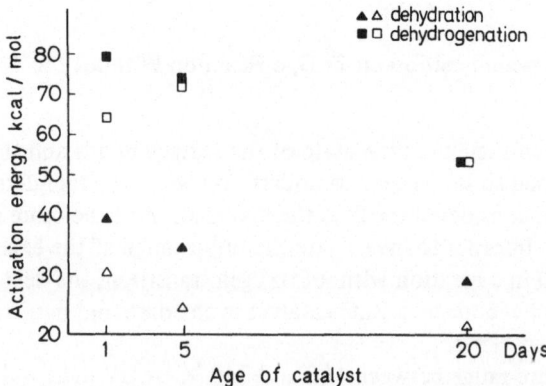

Fig. 13. Influence of UV irradiation and age of ZnO catalyst on activation energies of isopropanol decomposition. Open symbols, light; closed symbols, dark

146

In both reaction pathways the decomposition is going through a transition state, involving formation of two hydrogen bridges to adjacent oxygen and zinc atoms of the ZnO surface [96]. Between the surface oxygen atom and the oxygen atoms of the alcohol, a cationic H bridge is formed. Between the surface zinc and a C atom of the isopropyl alcohol, an anionic bridge is formed, as in a hydride. From the steric point of view, H bridges to the more distant O atoms of the surface alone would be more favourable. As a consequence, a cationic H bridge must be formed between an O atom of the surface and a C atom of the isopropyl alcohol. The electronegativity of the carbon is too low to form such a cationic H bridge. Thus we assume the transition states shown in Fig. 14. They enable a cyclic electron shift, the hydrogen atoms passing simultaneously from the alcohol bond into a bond with the surface atoms. As bent H bridges are necessary for the formation of acetone, the activation energy of the formation of acetone is observed to be higher than the activation energy of the formation of propylene.

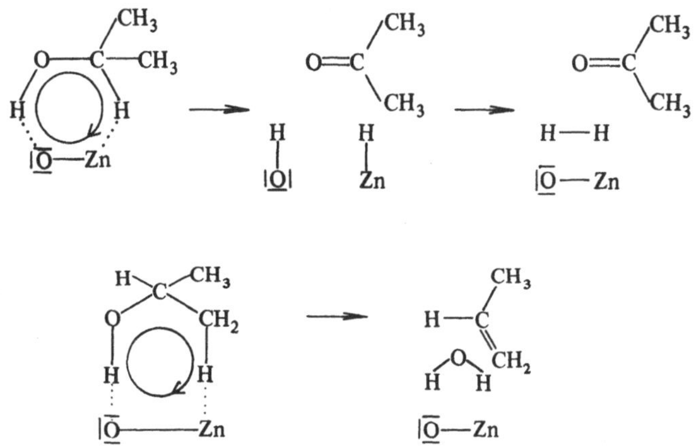

Fig. 14. Transition states with two hydrogen bridges for dehydrogenation (top) and dehydration (bottom)

When the catalyst is illuminated, the ZnO bonds are excited and loosened. In the linear combination of the ZnO bonds (Eq. 25) the nonbonding form increases at the expense of the bonding forms. This means that there is a shift of the bond electrons from the oxygen to the zinc atom. The electron density in the orbitals of the Zn atoms is increased, while the electron density in the orbitals of the oxygen atoms is decreased as compared with that of the ground state of the bond. A similar state is generated by the oxygen desorption of the surface where electron-enriched zinc atoms remain. ZnO, as an n-conductor, even in the ground state is furnished with more electrons than binding orbitals. Accordingly, the superfluous electrons occupy nonbonding orbitals thereby causing a loosening of the bonds. The formation of H bridges of the transition state, therefore, is more difficult; the activation energy should be increased (Fig. 15). Apparently, in the photoreaction an inversion of the atomic arrangement of the transition state takes place: the cationic H bridge is formed between the alcoholic oxygen of isopropyl alcohol and the zinc atom, now electron-enriched, of the excited surface bond. The anionic H bridge is formed between the C atom of isopropyl alcohol and the electron-exhausted oxygen of the excited surface bond. Fig. 15 shows the changes in the potential valleys as a consequence of the bond excitation. We see that the electron movement and the hydrogen shift need less energy as the character of the H bridges is inverted.

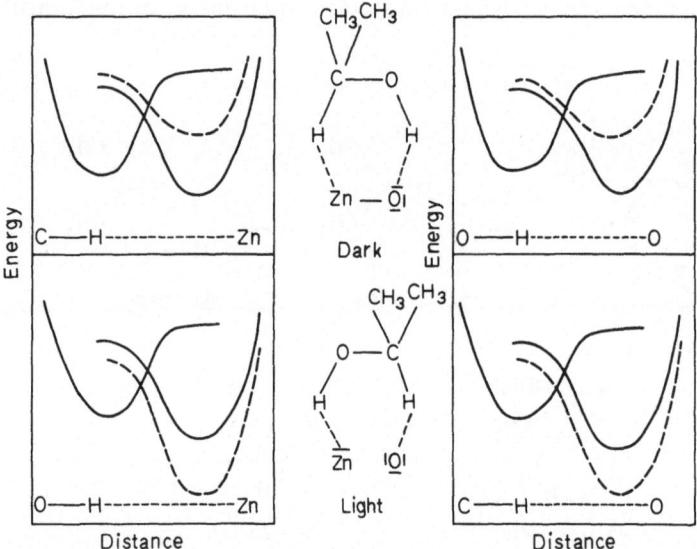

Fig. 15. Decrease of activation energy by excitation of the ZnO surface bond by illumination for the two transition states at the surface

Even reactions without oxygen transition demonstrate the important influence of the strength and the excitation state of the surface bonds of the catalyst on the amount of activation energy of the catalytic reaction.

5.6. Photocatalytic Oxidation Reactions on Metals

In catalytic reactions on metals a decrease in activation energy under UV illumination is also observed [97-99]. In experiments on CO oxidation on evaporated Ag, Au, Pd and Pt films [98], a decrease is observed in the activation energy due to illumination (Table 3), while no change is found in the order of the reaction. Under the experimental conditions, the surface of the metals is saturated with oxygen [99-105] (zero order with respect to O_2). The influence of the light on all metals is exerted in the same direction.

Table 3. *Temperature range in °C, kinetics, activation energy in kcal/mole and pre-exponential factor k_0 in s^{-1} for the first order, in torr/s for the zero order reaction*

	Ag	Au	Pd	Pt
Temperature range	300−550	380−530	230−350	240−500
$-dp_{CO}/dt =$	$k\,p_{CO}$	$k\,p_{CO}$	k	$k\,p_{CO}$
Dark activation energy	14,6	28,0	34.0	21.0
Photoactivation energy	6,25	10,0	23,8	18,8
$\log k_0$, dark	2,3	5,5	12,5	4,8
$\log k_0$, light	0,5	0	9,0	3,5

In these experiments, too, it is possible, therefore, to interpret the decrease in the activation energy in the light as due to excitation and loosening of the bond between noble metal and oxygen at the surface. The chemisorption bond between the oxygen atom and the noble metal atom may be described by a resonance similar to the resonance of the semiconductor bond. The bond is excited and weakened by photon absorption. The oxygen transition from the excited chemisorption bond to the CO molecule requires less energy than in the dark.

This interpretation gains strong support from the behaviour of Au catalysts (Fig. 16). Since the activation energy of the reaction is related to the strength of the chemisorption bond of the oxygen, the activation energy of the reaction changes if the chemisorption bond of the oxygen species participating in the reaction is changed. The peculiar dependence of the activation energy on temperature can be explained by the arrangement in the gold surface.

Like rolled metal sheets and wires, the gold film − having undergone evaporation and having been treated in vacuo at 400 °C for two hours − shows a well-ordered surface with only a few dislocations. When the gold film is heated to temperatures of about 500 °C, a large number of new dislocations appears, where catalytic reaction is likely to occur. On sheets and wires, too, the recrystallization beginning at 500 °C generates new lattice disturbances in the surface. We therefore assume that reaction on gold films not heated to temperatures above 500 °C likewise occurs on the whole surface area or, more precisely, on a uniform configuration of chemisorbed oxygen covering, if not the whole, at least a great deal of the gold surface. A high activation energy is necessary, but there is a high pre-exponential factor. Heating the gold to temperatures above 500 °C produces new dislocations. Hence, a new configuration of chemisorbed oxygen is generated where conditions favour reaction. Obviously, the number of dislocations is smaller than the number of normal surface atoms. Consequently, not only the activation energy but also the pre-exponential factor is diminished. Both the activation energy and the pre-exponential factor are smaller in the light than in the dark, because the number of oxygen atoms on the surface is smaller in the light than in the dark. As long as the number of successful collisions of CO molecules with the oxygen atoms of the surface increases with increasing temperature, we get a positive value for activation energy even in the light. When all oxygen atoms of the reactive chemisorption species participate in the reaction, it is no longer possible to increase the number of successful collisions by increasing temperature; the activation energy is zero.

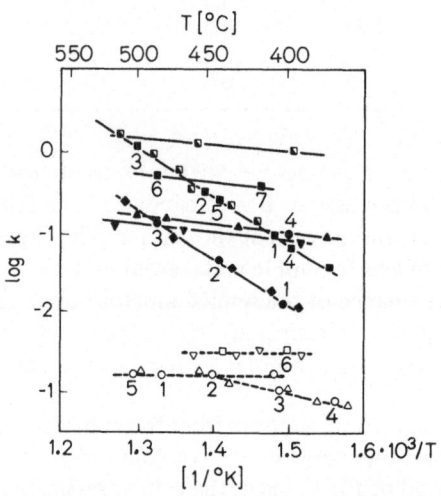

Fig. 16. Arrhenius plot for the CO oxidation on Au; top, dark; bottom, light. The numbers show the sequence of reaction temperatures. Between the points ■ 5 and ■ 6, ● 3 and ● 4, and ○ 5 and ○ 6 the catalyst has been heated above 500 °C for more than 1 hour

Once the gold has been heated to 500 °C, the precentage of successful collisions per total number of adsorbed oxygen atoms increases compared to the percentage before heating, in other words, less oxygen is adsorbed. However, these centres are completely involved in the reaction at lower temperatures because of their inferior activation energy. The saturation (activation energy zero) on Au films, heated up to 500 °C and rich in dislocations, occurs at lower temperatures than on Au films not heated to 500 °C which have fewer dislocations (bottom of Fig. 16).

6. Summary

By irradiating a solid with photons, it is possible to alter in a well controlled way the occupation of the bond orbitals in the surface of the solid. This procedure enables us to influence surface migrations of atoms of the solid, sorption reactions and catalytic reactions, and to reveal the bond changes fundamental to any surface reaction. Several sorption processes have been investigated and may be interpreted to a certain extent. The interpretation of catalytic processes, however, is still too general. One problem in particular remains unsolved: whether the adsorption states studied so far are identical with the unstable short-lived transition states which occur in catalytic reactions.

Acknowledgement

I thank the Deutsche Forschungsgemeinschaft, the Fonds der Chemischen Industrie and the Laboratory for Research on the Structure of Matter, Philadelphia, for sponsoring these investigations.

I thank my co-workers R. Barth, D. Binding, K. Hiltner and H. D. Müller for their continuous help.

I am indebted to Miss G. Kampf for revising the English manuscript.

7. References

1) Schwab, G.-M., Holz, G.: Z. Anorg. Allgem. Chem. *252*, 205 (1944).
2) – Trans. Faraday Soc. *42*, 689 (1946).
3) – Semiconductor Surface Physics (ed. R. H. Kingstone), p. 283. Philadelphia: University of Pennsylvania Press 1957.
4) – Festkörperprobleme (ed. F. Sauter) Vol. I, p. 188. Braunschweig: Vieweg & Sohn 1962.
5) Hauffe, K.: Advan. Catalysis *IX*, 187 (1957).
6) Garner, W. E.: Advan. Catalysis *IX*, 169 (1957).
7) Rees, A. L. G.: Chemistry of the Defect Solid State. London: Methuen 1954.
8) Garrett, C. G. B.: J. Chem. Phys. *3*, 966 (1960).
9) Stone, F. S.: Advan. Catalysis *XIII*, 1 (1962).
10) Bond, G. C.: Catalysis by Metals. New York: Academic Press 1964.
11) Wolkenstein, Th. Th.: Advan. Catalysis *XII*, 187 (1960).
12) Hauffe, K.: Reaktionen in und an festen Stoffen, p. 304. Berlin-Heidelberg-New York: Springer 1966.
13) Gorgoraki, V. I., Kasatkina, L. A., Levin, V. Yu.: Kinetika i Katalis *4*, 422 (1963).
14) Gobeli, G. W., Allen, F. G.: Intern. Conf. on Physics and Chemistry of Solids held at Brown University, Prepr. V.
15) – – Surface Sci. *2*, 402 (1964).

F. Steinbach

[16] Fischer, T. E.: Surface Sci. *13*, 30 (1969).
[17] Gobeli, G. W., Allen, F. G.: Phys. Rev. *137*, A 245 (1965).
[18] Cohen, M. L., Phillips, J. C.: Phys. Rev. *139*, A 912 (1965).
[19] Fischer, T. E.: Phys. Rev. Letters *21*, 31 (1968).
[20] – Helv. Phys. Acta *41*, 6/7 (1968).
[21] – Surface Sci. *10*, 399 (1968).
[22] Gerasimov, A. B., Dolidze, N. D., Kakhidze, N. G., Konovalenko, B. M., Chelidze, N.V.: Fiz. Tekh. Poluprov. *1*, 982 (1967).
[23] Landolt-Börnstein, Zahlenwerte und Funktionen, Vol. II, Part 6, p. 267. Berlin-Göttingen-Heidelberg: Springer 1959.
[24] Zizine, J., in: Radiat. Eff. Semicond. Proc. 1967 (ed. F. L. Vook), p. 186. New York: Plenum Press 1968.
[25] Mohanty, S. R., Nair, S. M. K.: Current Sci *37*, 283 (1968).
[26] Chon, H., Pajares, J.: J. Catalysis *14*, 257 (1969).
[27] – Prater, C. D.: Discussions Faraday Soc. *41*, 380 (1966).
[28] Gray, T. J. Amigues, P.: Surface Sci. *13*, 209 (1969).
[29] Urbach,F.: Sitzungsber. Akad. Wiss. Wien, Math.-Naturw. Klasse *139*, 353 (1930).
[30] Randall, J. T., Wilkins, M. F. H.: Proc. Roy. Soc. (London, Ser. B *184*, 390 (1945).
[31] Garlick, G. F. J.: Photoconductivity. In: Handb. Physik (ed. S. Flügge), Bd. 19. Berlin: Springer 1956.
[32] Bube, R. H.: Phys. Rev. *99*, 105 (1955); *101*, 1688 (1956); *106*, 703 (1957).
[33] Hauffe, K., Schmidt, R.: Physik und Chemie der Grenzflächen, Kolloquium der DFG in Marburg 123 (1969); – Photogr. Korr. *107*, 132 (1971).
[34] v. Baumbach, H. H., Wagner, C.: Z. Phys. Chem. (Leipzig) Abt. B *22*, 199 (1933).
[35] Heiland, G.: Z. Physik *142*, 415 (1955).
[36] – Discussions Faraday Soc. *28*, 168 (1959).
[37] Melnick, D. A.: J. Chem. Phys. *26*, 1136 (1957).
[38] Romero-Rossi, F., Stone, F. S.: Actes IIième Congrès International de Catalyse, Paris 1960, part II, 1481.
[39] Fujita, Y., Kwan, T.: Bull. Chem. Soc. Japan *31*, 379 (1958).
[40] Terenin, A., Solonitzin, Yn.: Discussions Faraday Soc. *28*, 28 (1959).
[41] Baru, V. G.: Probl. Kinetiki i Kataliza, Akad. Nauk SSSR *12*, 152 (1968).
[42] Lyashenko, L. V., Gorokhovatskij, Ya. B.: Dokl. Akad. Nauk SSSR *186*, 1125 (1969).
[43] Lisachenko, A. A., Vilesov, F. I.: Dokl. Akad. Nauk SSSR *176*, 1103 (1968).
[44] Wolkenstein, Th. Th., Baru, V.: Usp. Khim. *37*, 1685 (1968) (Russ. Chem. Rev. *37*, 724 (1968)).
[45] – – Surface Sci. *13*, 294 (1969).
[46] Ruppel, W., Gerritsen, H. J., Rose, A.: Helv. Phys. Acta *30*, 495 (1957).
[47] H. G. Fitzky: Phot. Korr. *103*, 173 (1967).
[48] Sancier, K. M.: J. Catalysis *5*, 314 (1966).
[49] Boreskov, G. K.: Advan. Catalysis *XV*, 285 (1964).
[50] Haber, J.: Zesz. Nauk. Akad. Gorn.-Hutn. Krakowie *7*, 13 (1967); C. A. *69*, 89913 (1968).
[51] Lyasshenko, L. V., Gorokhovatskij, Ya. B.: Kinetika i Kataliz *9*, 1180 (1968).
[52] Saltsburg, H.: U. S. Govt. Res. Develop. Rept. *68*, 109 (1968).
[53] Wolkenstein, Th. Th., Karpenko, I. V.: J. Appl. Phys. *33*, 460 (1962).
[54] – – Dokl. Akad. Nauk SSSR *165*, 1101 (1965).
[55]   Fiz. Tverd. Tela *9*, 403 (1967).
[56] Segal, E., Teodorescu, M.: Rev. Roumaine Chim. *13*, 1440 (1968).
[57] Hasebe, H., Kikushima, K.: Nippon Kagaku Zasshi *90*, 112 (1969); C. A. *70*, 71821z (1969).

152

[58] Munuera, G., Gonzalez, F., Machinabeitia, G.: An. Real. Soc. Espan. Fis. Quim., Ser. B 63, 763 (1967).

[59] – – Rev. Chim. Minerale 4, 207 (1967).

[60] Che, M., Naccache, C., Imelik, B.: J. Chim. Phys. 65, 1301 (1968).

[61] Fukuzawa, S., Sancier, K. M., Kwan, T.: J. Catalysis 11, 364 (1968).

[62] Landolt-Börnstein, Zahlenwerte und Funktionen, Vol. II, Part 6, p. 259. Berlin-Göttingen-Heidelberg: Springer 1959.

[63] Zhabrova, G. M., Vladimirova, V. I., Kadenatsi, B. M.: J. Catalysis 12, 238 (1968).

[64] Weis, L.D., Evans, T. R., Leermakers, P. A.: J. Am. Chem. Soc. 90, 6109 (1968).

[65] Wennerberg, A. N.: U.S. Patent 3.360.449 (1967); C.A. 68, 77942v (1968).

[66] Romero-Rossi, F., Stone, F. S.: Actes IIième Congrès Internationale de Catalyse, Paris 1960, 1481.

[67] Lyashenko, L. V., Gorokhovatskii, Ya. B.: Dokl. Akad. Nauk SSSR 186, 1125 (1969).

[68] Schwab, G.-M., Steinbach, F., Noller, H., Venugopalan, M.: Z. Naturforsch. 19a, 45 (1964)

[69] Dörffler, W., Hauffe, K.: J. Catalysis 3, 171 (1964).

[70] Lyashenko, L. V., Gorokhovatskii, Ya. B.: Teor. i Eksperim. Khim., Akad. Nauk Ukr. SSR 3, 218 (1967).

[71] Hauffe, K., Pusch, H.: Ber. Bunsenges. Physik. Chem. 72, 669 (1968).

[72] Binding, D., Steinbach, F.: Nature 227, 832 (1970).

[73] Steinbach, F.: Dechema-Berichte (1970) in the press.

[74] Steinbach, F., Hiltner, K.: Unpublished results.

[75] Fujishima, A., Honda, K., Kikuchi, Sh.: Kogyo Kagaku Zasshi 72, 10 (1969).

[76] Formenti, M., Juillet, F., Teichner, S. J.: C. R. Acad. Sci. Paris 270, 138 (1970).

[77] Boehm, H.-P., Kaluza, U.: J. Catalysis, in the press; – Physik und Chemie der Grenzflächen, Kolloquium der DFG in Marburg 134 (1969).

[78] Harkins, C. G., Shang, W. W., Leland, T. W.: J. Phys. Chem. 73, 130 (1969).

[79] Nelson, R. L., Teuch, A. J., Harmsworth, B. J.: Trans Faraday Soc. 63, 1427 (1967).

[80] Lunsford, J. H.: J. Phys. Chem. 68, 2312 (1964).

[81] Gomes, W. P., Freund, T., Morrison, S. R.: Surface Sci. 13, 201 (1968).

[82] Steinbach, F.: Nature 215, 152 (1967).

[83] – Krieger, K. A.: Z. Phys. Chem. N.F. 58, 290 (1968).

[84] – Z. Phys. Chem. N.F. 60, 126 (1968).

[85] – Z. Phys. Chem. N.F. 61, 235 (1968).

[86] – Nature 221, 657 (1969).

[87] – Symposium on Electronic Phenomena in Chemisorption and Catalysis on Semiconducters (eds. K. Hauffe, Th. Th. Wolkenstein), p. 146. Berlin: Walter de Gruyter 1969.

[88] – Z. Phys. Chem. N.F. 71, 14 (1970).

[88] Steinbach, F.: Z. Phys. Chem. N.F. 71, 14 (1970).

[89] Schwab, G.-M., Siegert, R.: Z. Phys. Chem. N.F. 50, 191 (1966).

[90] – Derleth, H.: Z. Phys. Chem. N.F. 53, 1 (1967).

[91] – Surface Sci. 13, 198 (1969).

[92] Steinbach, F., Barth, R.: Ber. Bunsenges. Physik. Chem. 73, 884 (1969).

[93] Müller, H. D., Steinbach, F.: Nature 225, 728 (1970).

[94] Steinbach, F., Müller, H. D.: Ber. Bunsenges. Physik. Chem. 74, 935 (1970).

[95] – Chem. Ing.-Tech. 42, 1065 (1970).

[96] Müller, H. D.: Diplomarbeit, p. 94, Universität München 1969.

[97] Moesta, H., Breuer, H. D., Trappen, N.: Ber. Bunsenges. Physik. Chem. 73, 879 (1969).

[98] Steinbach, F.: Z. Phys. Chem. N.F. 71, 29 (1970).

[99] Baddour, R. F., Modell, M.: J. Phys. Chem. 74, 1392 (1970).

[100] Lewis, R., Gomer, R.: Surface Sci. 12, 157 (1968).

[101] Morgan, A. E., Somorjai, G. A.: Surface Sci. 12, 405 (1968).

F. Steinbach

[102] Imre, L.: Ber. Bunsenges. Physik. Chem. *72*, 863 (1968).
[103] Murgulescu, I. G., Vass, M. I.: Rev. Roumaine Chim. *13*, 373 (1968).
[104] Bailitis, E.: Z. Phys. Chem. N.F. *64*, 302 (1969).
[105] Ertl, G., Rau, P.: Surface Sci. *15*, 443 (1969).

Received December 7, 1970

SPRINGER - VERLAG
BERLIN · HEIDELBERG · NEW YORK

Fortschritte der chemischen Forschung
Topics in Current Chemistry

Schriftleitung: F. Boschke

Band 16, Heft 1:

With 8 tables
IV, 102 pages (26 p.
in German). 1970
Soft cover DM 54,–

Reactive Intermediates

Abramovitsch/Sutherland: Nitrenes are reactive intermediates having a monovalent nitrogen atom with a sextet of electrons in its outer shell. Heaney demonstrates that the chemistry of benzyne to be worthy of study. Winterfeldt: A review about a new development in sigmatrope reactions (generation and recombination of two allyl-radicales).

Band 16, Heft 2:

NMR Spectroscopy of Annulenes

By R. C. Haddon and L. M. Jackman

With 6 figures
III, 118 pages. 1971
Soft cover DM 64,–

This is a discussion of the NMR properties of annulenes (and their dehydro equivalents), the relevant anions and cations, systems with bridges and certain related systems. Special attention is given to the qualitative analysis of aromaticity and anti-aromaticity.

Organometallic Compounds in Industry

Band 16, Heft 3/4:

With 16 figures
III, 183 pages (104 p.
in German). 1971
Soft cover DM 64,–

Contents: R. F. Heck, Addition-Elimination Reactions of Palladium Compounds with Olefins. F. W. Frey and H. Shapiro, Commercial Organolead Compounds. A. Gumboldt, Metallorganische Verbindungen als Katalysatoren der Olefin-Polymerisation. H. Weber, Metallorganische Verbindungen als Katalysatoren zur Herstellung von Stereokautschuken. A. Bokranz und H. Plum, Technische Herstellung und Verwendung von Organozinnverbindungen.

In kritischen Übersichten werden in dieser Reihe Stand und Entwicklung aktueller chemischer Forschungsgebiete beschrieben. Sie wendet sich an alle Chemiker in Forschung und Industrie, die am Fortschritt ihrer Wissenschaft teilhaben wollen.

In der Regel werden nur Beiträge veröffentlicht, die ausdrücklich angefordert worden sind. Schriftleitung und Herausgeber sind aber für ergänzende Anregungen und Hinweise jederzeit dankbar. Manuskripte können in den „Fortschritten der chemischen Forschung" in Deutsch oder Englisch veröffentlicht werden.

Jeder Band der Reihe ist einzeln käuflich.

This series presents critical reviews of the present position and future trends in modern chemical research. It is addressed to all research and industrial chemists who wish to keep abreast of advances in their subject.

As a rule, contributions are specially commissioned. The editors and publishers will, however, always be pleased to receive suggestions and supplementary information. Papers are accepted for "Topics in Current Chemistry" in either German or English.

Any volume of the series may be purchased separately.